In memory of Roger Chapman

THE DIVE

STEPHEN MCGINTY

THE DIVE

THE UNTOLD STORY OF THE WORLDS
DEEPEST SUBMARINE RESCUE

HarperCollins*Publishers*

HarperCollins*Publishers*
1 London Bridge Street
London SE1 9GF

www.harpercollins.co.uk

HarperCollins*Publishers*
1st Floor, Watermarque Building, Ringsend Road
Dublin 4, Ireland

First published by HarperCollins*Publishers* 2021
This edition published 2022

1 3 5 7 9 10 8 6 4 2

Text © Stephen McGinty 2021

Images courtesy of: p.5, 72 Al Trice; p.9 RTÉ Archives; p.188 Dipper
Historic/Alamy Stock Photo; p.266 Naval Information Warfare Center
Pacific; p.304 PA Images/Alamy Stock Photo; p.310 PJF Military
Collection/Alamy Stock Photo; p.313 Harry Dempster/*Daily Express*
(press handout); p.317 PA Images/Alamy Stock Photo

Stephen McGinty asserts the moral right to
be identified as the author of this work

A catalogue record of this book is
available from the British Library

ISBN 978-0-00-841078-0

Printed and bound in the UK using 100%
renewable electricity at CPI Group (UK) Ltd

MIX
Paper from
responsible sources
FSC™ C007454

This book is produced from independently certified FSC™ paper
to ensure responsible forest management.

For more information visit: www.harpercollins.co.uk/green

To LA, from the depths of my heart

'Out of the depths have I cried unto thee, O Lord.'

PSALM 130, SONG OF ASCENTS

CONTENTS

CONTENTS

CAST

BRITISH TEAM – VICKERS OCEANICS

Roger Mallinson, *Pisces III* pilot
Roger Chapman, *Pisces III* co-pilot
Sir Leonard Redshaw, chairman of Vickers Shipbuilding
Greg Mott, managing director of Vickers Oceanics
Commander Peter Messervy, general manager of Vickers Oceanics
Bob Eastaugh, operations manager
Len Edwards, master of the Vickers *Voyager*
Ralph Henderson, surface officer
David Mayo, diver and communications
Desmond 'Des' D'Arcy, chief submarine pilot
Geoff Hall, electrical engineer
Doug Huntington, project engineer
Ted Carter, assistant technical manager, mechanical
Maurice Byham, sales engineer
Bob Hanley, surface officer and diver
Harold Pass, technical manager
Dick Nesbitt, electronics engineer
Terry Storey, assistant technical manager

Roy Browne, pilot
Mike Bond, diver

CANADIAN TEAM – HYCO

Dick Oldaker, president
Jim McFarlane, operations manager
Al Trice, co-founder
Mike Macdonald, senior pilot
Jim McBeth, *Pisces* project engineer
Bob Starr, pilot
Bob Holland, pilot
Al Witcombe, pilot
Steve Johnson, pilot

US TEAM

Commander Ramos, US Navy liaison officer
Commander Earl Lawrence, US Navy salvage master
Commander Bob Moss, US Navy deputy supervisor of
    salvage
Bob Watts, CURV-III programme manager
Larry Brady, principal pilot
Tom Wojewski, sonar technician
Denny Holstein, deck supervisor
John De Friest, mechanical technician
William Sanderson, electronics technician
William Patterson, photo technician

SURFACE SHIPS

Vickers *Voyager*
HMS *Hecate*
HMS *Sir Tristram*
John Cabot, Canadian Coast Guard icebreaker and cable-
  laying vessel
USS *Aeolus*, US Navy cable-laying vessel

SUBMERSIBLES

Pisces II
Pisces III
Pisces V
CURV-III (cable-controlled undersea recovery vehicle)

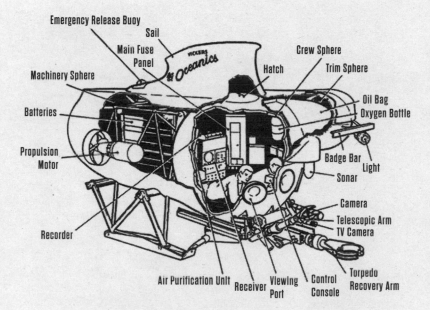

Emergency Release Buoy

Sail

Main Fuse Panel

Machinery Sphere

Batteries

Propulsion Motor

Recorder

VICKERS Oceanics

Crew Sphere

Trim Sphere

Hatch

Oil Bag

Oxygen Bottle

Badge Bar

Light

Sonar

Camera

Telescopic Arm

TV Camera

Torpedo Recovery Arm

Air Purification Unit

Receiver

Viewing Port

Control Console

PROLOGUE

In the cabin the gentle rock and roll of the ship is as good as a lullaby to Roger Chapman. A lifetime at sea has taught him to sleep whenever possible and grab what you can. He might only have turned in a couple of hours earlier, but this is all the rest he needs, and when the alarm goes off at 1 am he rises, washes himself in the basin and dresses in a pair of blue jeans and a grey shirt, then pulls on a pair of old blue overalls. He thinks for a second about picking up a heavier jumper, but he's only going to be gone a few hours so he doesn't see the point. Before going out of the cabin, he leaves the pilot's log open, ready for his return. A couple of minutes later he steps out of the cabin, swings by the canteen where the chef has already prepared their packed lunches and heads up on deck. He likes the quiet of the early hours. Although the Vickers *Voyager*, the company's 2,850-ton command vessel – red with a white trim, and capable of carrying and servicing two submersibles – is operating around the clock, at night there are fewer people and a sense of stealthy calm settles over the ship.

On deck he feels the mild chill of the evening breeze, then looks down at the dark waters of the Atlantic Ocean. The nearest land is over 150 miles distant. The immense blackness under the horizon is illuminated only slightly by the moon, trapped behind a curtain of grey clouds. It's time to go to work. He walks to the internal hold, where the white shape of *Pisces III*, their mini-submersible, is sitting, and where his colleague Roger Mallinson has clearly spent the last few hours. A skilled engineer, who in his free time likes nothing more than building miniature steam engines, Mallinson has been frustrated at the poor performance of the submarine's manipulator arm, the five-foot-long extendable metal talon, so he has decided to strip it down and rebuild it.

Mallinson, a tall, thin man with a bushy brown beard and piercing blue eyes, is not having a good day, or evening. He's feeling rough. He has not slept for 24 hours and has barely eaten. The memory of a cold meat and potato pie he'd been served from a peeling Formica counter at a down-at-heel pub near the airport, his last real 'meal', refuses to fade. He might be coming down with food poisoning and has been complaining about having a touch of the runs, which he hopes is passing.

When the manipulator arm is acting up, the issue is either mechanical or hydraulic, but on this occasion it's both. Then there's the matter of the oxygen supply. The sub has around half a bottle, more than sufficient to do a 10- to 12-hour job, but Mallinson stops to think for a second and decides to replace it with a full tank. He gets up, climbs out of the sub with the oxygen tank and heads off to the oxygen store, three decks down. Luckily it's open, and although he knows he should get permission from Ralph Henderson, the field officer, he doesn't want to approach him, not after the way Henderson

Pisces V being loaded onto Vickers *Voyager* at Cork Harbour.

spoke to him earlier in the week. (Mallinson had some concerns about the aft sphere hatch and asked for repairs prior to his last break. When he got back on board he asked Henderson if they'd been carried out, and his boss was less than polite in his reply. They hadn't, Mallinson was told, and if he had a problem with that, he didn't need to dive.) Mallinson pulls out a full tank – replacing it with the half-full tank from the sub, even though he shouldn't – hoists it over his shoulder and clambers back up the stairs.

Chapman finds Mallinson where he'd left him: behind the controls and grappling with the final repairs to the manipulator arm. Mallinson insists he has snatched a few minutes' sleep but Chapman is doubtful, and now there's no time left for even a quick nap. They have a schedule to maintain and it's almost 1.30 am. Chapman climbs down into the sphere, carrying the sandwiches, a flask of coffee, a small carton of milk and a Tupperware box of sugar.

Pisces III is connected to winches and wire restraints in *Voyager*'s ceiling that enable her 12-ton weight to be moved easily towards the stern, where steel strops abruptly halt the craft. The front of *Pisces III* is now directly over the sea, with the portholes offering a view of the dark wash of waves below.

Chapman and Mallinson, sitting in the sub's tight operational sphere, just six feet across, begin to work through the pre-dive checklist for what on their paperwork is designated as 'Dive No. 325'.

All equipment secured: check
Ports cleaned: check
Emergency buoy release working: check

Emergency buoy line length equates to possible maximum depth:
  check
Main hatch cleaned and greased: check
Battery oil topped up at: 00.45 hrs
Oxygen main: 3,000 psi oxygen reserve: 2,900 psi
Cabin barometer set: check
Clock wound and set: check
CO_2 Dräger kit: check
Drop-weight wrench: check
TV system functioning: check
Jettison & emergency buoy valves shut: check
Cabin vent valve shut: check
Sonar: check
Air scrubber: check
External lights: check
VHF radio: check
Operate emergency trip and confirm: check
12-volt emergency battery power: check
Prop motors: check

The final item on the checklist is the operational efficiency of
the manipulator and claw, an extendable mechanical arm that
allows the crew to pick up and move around equipment,
which is crucial for their upcoming mission. Mallinson tenses
a fraction as Chapman tests it out. Outside the crew on deck
watch as it swivels and turns, the hand moving fluidly in and
out. A thumbs-up is enough to secure the final 'check'. *Pisces
III* is good to go.

As they enjoy a moment of ease, Mallinson is thinking
about the previous day: the day of the dolphins. He had been

working on the underwater telephone in the comms room when suddenly a pod of dolphins appeared on the line, squeaking and chortling on the same frequency. Mallinson had 'spoken' to them before and knew how loud they could be. On the bottom you could lose messages from the top because of their incessant conversation, but as he explained to his shipmates, 'I don't mind losing messages to dolphins.' When a crew member stuck his head around the door and said a large pod were off the bow, Mallinson asked him to mind the comms while he fetched his cine camera and went up on deck. 'I'd never seen anything like it, the whole sea as far as you could see, horizon to horizon, was dolphins.' Yet by the time he had the camera out of the case, all he caught on film were six tails disappearing into the sea. In the sub, Mallinson wonders if they will return.

The weather at 1.30 am is relatively calm, but the wind is freshening and a storm is forecast. *Pisces III* is lowered over the side and into the sea under the careful watch of a pair of divers on a Gemini RIB (rigid inflatable boat). Once one of them has scrambled on top of *Pisces III* and disconnected the winch, the controller contacts Chapman and Mallinson on the VHF radio and instructs, 'Clear to vent.'

On these words, *Pisces III* expels the air from her buoyancy tanks and begins a controlled dive down through the first light-filled fathoms towards the inky darkness below. It will take 40 minutes to reach the bottom, where the intense pressure of the water is 800 psi (pounds per square inch), the equivalent of 50 tons across the entrance hatch.

Mallinson looks out of the porthole window and steers by the direction of what he describes as crud, little flecks of particles moving with the current. If the crud is moving up, you're going down. If the crud is going down, you're going up. It can be noisy in the sphere but Mallinson has his headphones on, listening to the echo sounder telling him where the bottom is and the cable lies. Then there's the noise of the underwater telephone and the gyro singing away. At times it's so loud you can hardly hear yourself think.

Pisces III's powerful thallium light beams – 1,000 watts – are spotlights that attract a colourful crowd to the sandy stage. At 1,700 feet cod, haddock, skate and the occasional eel are likely to sashay by, and Chapman never tires of the sight, never takes for granted the three portholes, their four-inch panes of glass, and the wonders each one opens up. The years spent as a naval submariner sealed into a metal tube, blind to the underwater world through which he and his comrades silently glided, have sharpened his senses, his appetite for actuality. Lying in his bunk, he often wondered what exactly lay beyond the nuclear submarine's steel shell, or where in the world they actually were, but now on each new dive he can see.

In the early hours of this Wednesday morning what he can see is a seabed of grey and brown, and sand grains slowly spiralling through the atmosphere. While Mallinson focuses on piloting *Pisces III* in the direction of the sonar pinger, a portable beacon whose repetitive soundwaves can be tracked to their destination, Chapman can enjoy the view. The Atlantic was clean, thought Chapman, unlike the North Sea with its sunken clutter of discarded pipe work, flanges and beer cans. There were times on previous dives for the oil industry that he'd thought the seabed more closely resembled a garbage tip: an embarrassment of detritus hurled into the depths.

Vickers *Voyager* and *Pisces III* have been hired by the British Post Office for the final stage of a £30 million project that is now entering its fourth year. The Post Office and the Canadian Overseas Telecommunications Corporation have laid a cable from Widemouth Bay in Cornwall that stretches 3,250 miles across the Atlantic to Beaver Harbour in Nova Scotia. The CANTAT-2 cable, wrapped in an armour coating, is 2⅜-inch diameter and able to carry 1,840 separate transatlantic telephone calls simultaneously, more than the combined capacity of all the other transatlantic cables in operation. The project has been three and a half years in development, with cable laying taking six months as the icebreaker *John Cabot* slowly unspooled this 'thread across the ocean'. Now, however, it's in danger of being ripped up by deep sea trawlers, whose nets are cast out to a depth of one mile. While *Voyager* and *Pisces III* headed west burying the cable as they progressed, a few hundred yards a day, on the other side of the Atlantic, heading east was another Vickers ship carrying her sister submarine *Pisces V*, which was doing the same job in the opposite direction.

Mallinson is zig-zagging *Pisces III* from left to right as the sub pursues the signal from the pinger, an 18-inch sonic device, which, like a baton in a relay race, is picked up by each new crew at the spot where the last crew had finished work for the day, or night. Mallinson will pick it up and later drop it down again at the end of their shift, as the pinger slowly moves along the length of the cable.

Chapman's contemplation of the deep is broken by the first

glimpse of the heavy dark cable, resting as yet unburied on the soft sand. The beam of *Pisces III*'s spotlight has found CANTAT-2 before Mallinson found the pinger. He tells Mallinson, who trims the sub's trajectory slightly, sending the sub skimming towards the job site. Once in place, Mallinson takes charge of the mud pump, a water jet powerful enough to displace the surface sand and mud to create a shallow grave into which the heavy armoured cable is then laid. They are travelling west on a slight downward gradient with the continent of North America 3,000 miles away in the dark distance. For the first hour Mallinson uses his left hand on two motor throttles to maintain *Pisces III* in position and his right to manipulate the mud pump nozzle and blast the seabed. He's surprised by the ease with which the sand parts and the new trench emerges. The sandy bed is fine, easily displaced and quite unlike previous sections of the job, which, stubbornly, had taken three or four passes to achieve the correct depth for the ditch.

The submarine has a small battery-powered stereo tape recorder and, given the sympathetic contours and acoustics of the vessel, the sound is, as Mallinson describes to friends, 'quite sensational'. Yet the musical accompaniment he favours to his underwater manoeuvres is not popular with everyone. In a year that has already seen the release of Elton John's *Goodbye Yellow Brick Road*, David Bowie's *Aladdin Sane*, Led Zeppelin's *Houses of the Holy* and, what will become many people's soundtrack to the decade, Pink Floyd's *Dark Side of the Moon*, Mallinson's choice antedates them all by almost 250 years. He prefers to work to the soothing sound of organ music by Johann Sebastian Bach.

At 4 am the pair stop for a short rest, and break out the coffee flask and sandwiches – cheese and chutney for Chapman, strawberry jam for Mallinson. Then the pair swap positions, with Chapman taking the controls. Slowly the hours tick by, while on the surface night gives way to dawn. It's just after 6 am, around the time their batteries begin to dim, that they spot a problem. The next section of Post Office cable is not lying on the seabed but appears to be hovering – floating – three feet above the surface. *Pisces III* moves further west along the cable and her spotlights cast a bright light on what's wrong. The sub has now followed the cable down to a little over 1,700 feet where there's an obstruction the surveyors had not anticipated. Ahead lies a trough measuring around 25 feet wide, across which the cable is suspended like a trapeze wire. On the other side the seabed rises higher still. The trench and suspended cable are perfectly positioned for a passing trawler's hooks and nets to snag on. Although far off the coast of Ireland, both men know that 'Murphy's Law' operates in international waters.

Pisces III moves back and forth along the trench, filming the extent of the crevasse and explaining the issue to the surface team. The decision is made to make a start at cutting away at the eastern bank with the manipulator nozzle, to shave down the slope and allow the pipe to sit flush along the bottom. For 40 minutes they sand it down. Then they film the extent of their work so far and drop the audio pinger to mark their position, enabling the next shift to continue their work.

Chapman groans, 'This will add days to the job,' while Mallinson is concerned that they have lingered a little too long on the task at hand, wearing down their battery power. Privately he thinks they've been rather stupid.

On the surface, *Voyager* begins to make preparations for *Pisces III*'s ascent. The two-man team of divers dressed up in wetsuits, fins, masks and air tanks climb down the rope ladder and into the Gemini rescue RIB, then head out towards the bright orange buoy that bobs on the surface, marking *Pisces III*'s location. Even at a depth of 1,700 feet, Chapman and Mallinson can hear the soft phuft-phuft of the Gemini's outboard engine over the undersea microphone, an audible sign that permission to ascend is imminent.

When the call comes through a few minutes later, Mallinson begins pumping oil into the submarine's ballast bags, which quickly begin to shift the vessel's buoyancy from negative to positive. *Pisces III*'s skids then softly lift off from the sandy seabed and the sub begins to slowly rise out of the darkness towards the light. Chapman, a stickler for neatness, clears away the last of their sandwiches, then screws the flask of coffee good and tight. As they rise towards the surface, Mallinson notices a change in the direction of the crud and so the tide – it's now running east.

At 9.17 am the interior of *Pisces III* begins to lighten a little as the sub arrives into the shallow, sun-filled fathoms and the portholes offer a clearer view of a pair of passing cod.

'Welcome back,' shouts Ralph Henderson, the operations controller in charge of the recovery, over the shortwave VHF radio. Henderson is standing on the port bridge wing of *Voyager*, with a clear view of the Gemini and the rescue diver David Mayo.

At 9.18 *Pisces III* breaks the surface, where she begins to bob and buffet as the choppy waters slap against her hull. Since the portholes are on the bottom of the craft, neither Chapman nor Mallinson can yet see what's going on, but they can hear the clatter as the Gemini bangs up against the sub and Mayo clambers on top.

Ernie Foggin is running the Gemini this morning. Known as 'Uncle Ernie' to the younger crew members, his cabin door is always open to those who are homesick or lovelorn, or just in need of a Woodbine and some wise words. As the sub has been rising, he's been hauling in the sub's marker buoy and almost 2,000 feet of rope. Over many missions, trial and error have taught him and the rest of the team not to try to neatly coil the rope – there isn't the time – but just to dump it in the steel bin kept in the Gemini. Hurl it in randomly and it comes out just fine, but try to control the coil and it all knots up.

On *Pisces III*, Mayo, who will later describe the sea conditions as 'confused and large', knows his job. First he disconnects the long guide rope that ties the sub to the surface buoy and is used to give a permanent visual fix on her submerged position. The next task is to connect the tow line from *Voyager* to *Pisces III*, which means putting the tow line's snap hook into the spliced eye of a four-foot pendant, which is attached to the lower towing shackle fixed on the aft sphere of the sub.

He is passed the end of the tow line by Foggin, and, holding the snap hook in his right hand and the eye of the pendant in his left, he connects the two. He notices at the time that there's a lot of slack in the rope. He then moves up the port side of the sub and up towards the sail, where he would normally sit for the tow in.

'Tow line connected,' shouts Foggin over the VHF radio. The sea is rough and a wave pushes the Gemini away from *Pisces III*.

Mayo then lets go of the tow rope and pendant, aware that the tow rope is running over his left shoulder to his right-hand side and so clear of the aft hatch. But while passing the line over his head he feels a pull on the rope, 'as if the weight of the tow was coming onto it'.

He then carefully clambers onto the port side motor and then up into the sail. He's now facing forward over the starboard bow. When he turns around to adjust his position, he sees that the towing rope is around the hexagonal after-hatch securing bolt, running in an anticlockwise direction, the exact direction in which the bolt requires to be turned to loosen. He can also see that there's weight on the line.

Mayo tries to signal the danger. He can't contact Foggin in the Gemini as the little boat is out of sight, lost in the trough of a wave, so he signals with his hands to *Voyager*.

Henderson, who has now moved down to the port sponson deck on *Voyager*, can already see that the tow line has been washed over the side of *Pisces III* and is now lying across her aft deck. He orders the bosun who is operating the diesel-powered winch to pay out the tow line, thinking it will give Mayo enough slack to clear it from the stern of the sub. Just at this moment, one of the onboard staff from the British Post Office is on deck taking pictures of the recovery.

Mayo then scrambles out off the sail and moves down to the aft sphere to free the line. He has only managed to get one leg out and over the sail when the stern of *Pisces III* suddenly dips. At first he thinks this wash of waves will clear the line of its own accord, but he's horrified to see the aft hatch has flipped open and is now standing at an angle of 20 to 30°, with water now beginning to rush in. The tension of the tow rope against the hexagonal bolt has loosened it enough to open it, and the hatch is quickly swept away.

Henderson has looked away to give the order, and when he looks back he can see the aft hatch is off. He shouts to the bosun to secure the tow line around the capstan for extra support, then contacts the bridge to tell them to kick hard astern to take way off the ship and so slow the vessel down. They start to pull in the tow line, but when *Pisces III* begins to sink the ship's engines are stopped. Henderson can feel the sickly blossom of anxiety and adrenaline ignite across his chest, but not enough to diminish his sense of control and command. Where there's a problem, there's always a solution.

Inside the sub, Chapman and Mallinson are soaked in sweat from the humidity that builds up inside the sphere, and are now hungry for a late breakfast of bacon and eggs. When the water alarm sounds, neither man panics or even displays the mildest sense of concern. It's common for the condensation that builds up in the sub's aft sphere, a smaller circular storage compartment where oil and equipment are kept, to trigger the alarm. It has happened a couple of times before.

As well as the sound of the water alarm there's a second continuous rasping ring, but it still takes three seconds for both men to realise something is wrong. At first they think it might be an electrical fault triggering a false alarm, but then the frantic shouting begins to come over the VHF radio.

'There must be condensation in the aft sphere,' shouts Chapman.

Then *Pisces III* tips backwards at a 45° angle, hurling Mallinson to the back of the sphere.

Over the radio they hear the words, 'The diver is indicating something.'

As soon as the words from outside on the surface echo around the sub, *Pisces III*'s stern takes an even sharper dip down. The sub quickly sinks beneath Mayo, and he watches as her white shape blends into the ocean's blue, then disappears completely.

Inside, Chapman shouts at Mallinson to look at the 'bloody depth gauge'.

The needle has already touched 100 feet.

As the depth gauge hits 175 feet, *Pisces III* comes to a juddering halt, which causes both men to lose their footing and clatter against the submarine's steel casing. Then the submarine begins to shake violently. As they struggle to secure themselves, they realise that *Pisces III*, instead of sinking level is now tilted face down, like a diver halted mid-plunge. A sickening swaying begins, as, pendulum like, *Pisces III* sweeps back and forth, then up and down. As Chapman will later recall, 'the non-stop violence of the movement' was like being 'a rat in a terrier's mouth'.

Chapman thinks about shouting to Mallinson to find out what the hell is going on. Then he realises. The nylon tow rope that connects *Pisces III* to *Voyager* has uncoiled to the end of the line, gone taut and halted – momentarily – the submarine's rapid descent. He knows the line cannot hold, not with *Pisces III* now flooded with sea water and a full ton heavier. The tow line has a breaking strength of six tons and *Pisces III* weighs more than twelve. The line will undoubtedly snap. The question is, when?

Seconds merge into minutes as, topside, the team begin to quickly figure out an emergency rescue operation. The decision is made to send a diver down almost 200 feet, carrying a heavier tow line, which, if successfully deployed, will ease the tension on the nylon line and stabilise the sub. The ship's crew in the well deck rush to fetch the spare tow line while the Gemini returns to the stern with the diver ready to take the line down to secure it onto *Pisces III*.

Yet when both men are told of the plan, they realise it's an impossible task; almost a suicide mission. The violent motion of the submarine will make achieving a safe purchase for the diver exceedingly difficult and dangerous, and then to fix a second line as *Pisces III* is yanked up and down as *Voyager* rides the crest of waves is madness. It will never work, but Mallinson and Chapman have their own immediate concerns.

The underwater telephone has a spare lead–acid battery. This has come loose, and it now begins to swing with all the menace of a breeze block and the added danger of an acid leak if it cracks open. Then there's the big sonar set that has also broken free and is bashing both men. Mallinson grabs hold of it, wrenches it off the cables and throws it into the back of the sub. The two electrical wires are left hanging in the air. The water alarm isn't just bleeping, it's 'screaming', so Mallinson, when he finds it amid the confusion, shuts it down. Both men know full well they're in trouble and no longer need an alarm to tell them.

The question then becomes, what can they do to lighten their load? Mallinson remembers the sub's 400 lb drop weight, for use in the event of an emergency, which renders the submarine unable to blow water out of the ballast tanks. It's a large round lead block attached to the sub's base that can be jettisoned to lighten the vessel and so assist an ascent. The weight is fixed to *Pisces III* by means of a rigid steel bolt on the floor of the sub, but finding the spanner as the sub continues to spin is difficult. Mallinson can feel his pulse begin to rise and his breathing race as he scours the sub for the tool, and he concentrates on blocking out the surrounding din. *Pisces III* has become an echo chamber, with the microphone picking up the sound of the waves, the churn of the propellers above and voices from *Voyager* urging them to 'hang in' – as if they had any kind of choice. He needs to focus.

He finds the spanner and positions himself over the small square hole resembling a safe in the submarine's floor, with the bolt where the combination dial would be. He fits the box spanner around the bolt, and is about to put his weight behind a twist when the sub shifts. He bangs against the wall. He then moves back into position. The screw on the bolt has a long thread that requires between 25 to 30 turns before release, becoming stiffer and stiffer the further out the screw goes. One turn. Two turn. Yet with every turn, Mallinson is fighting to stay on his feet. Slowly the bolt begins to emerge out of the floor.

The final turn triggers an audible bang as the weight breaks free but there's an immediate echo: a second loud bang. The nylon tow rope has snapped. Its ability to bear the weight is finite, and when the break comes it's audible over the underwater microphone.

On *Voyager* the rams are brought in, ready to pass the new rope through the 'A' frame in order to give it a fair lead. It's then that the crew hear the tow line snap. They all look at each other as the rope line goes slack.

The bucking caused by the tow line is enough to induce nausea in both men. *Pisces III*, now no longer connected to *Voyager*, swiftly tilts to a 90° angle with the stern face down and begins to free fall. Although both men know this moment would come, it's still a shock as the sunlight from the surface begins to dim the deeper *Pisces III* drops. 'It was horrible falling down backwards,' recalls Mallinson.

The depth gauge reads 250 feet.

Mallinson picks up the underwater telephone and begins to read out their depth to *Voyager*.

'300 feet ... 350 feet ... 400 feet.'

Chapman starts to secure equipment ahead of impact. He knows that a loose component might not only break on impact but become a lethal projectile when ricocheting around a steel sphere. He disconnects the sonar set from its bracket, puts it at his feet and begins to go through the emergency procedures. All electrical power is switched off to minimise the risk of an explosive fire on impact, including the 12-, 24- and 120-volt systems. To brace for impact and prevent broken bones, Chapman stacks the cushioned seat covers towards the back of the sub where they will soon be thrown. He then turns to Mallinson and says, 'This is it, then.' The submarine is now in pitch darkness.

The sensation of falling is pronounced and audible. The

motors attached to the sub are powering through the reduction boxes and driving the electric motors at extreme speed. Mallinson thinks about trying to use the motors to slow the rate of descent but is fearful that they are spinning at such high revs that any interference now might burn them out or trigger a fire. The noise is intense, the motors 'screaming like a Stuka dive-bomber'.

Mallinson says nothing. Instead he continues to stare at the luminous depth gauge and its racing hand, and continues his count.

'800 feet.'

'900 feet.'

'1,000 feet'

Privately Chapman begins to fear that the sub might have drifted into deeper water, while also calculating her odds of surviving a sudden impact. They are roughly half a mile from a dip in the continental shelf, where the depth drops to 3,000 feet straight down.

'1,200 feet.'

'1,300 feet.'

'1,400 feet'.

Chapman half-states, half-asks, 'Must get there soon?'

Mallinson suddenly has a thought and shouts to Chapman, 'Bite on a rag.' So they quickly stuff their mouths and hope it will prevent them from biting clean through their tongues on impact. In the final seconds Mallinson remembers his father's warning about not going down in submarines. He tries to steel himself against what he knows will be a 'bloody great bang', the sudden rush of sea water and imminent death.

Pisces III strikes the seabed with a shuddering jolt that sends both men smashing against the sides and colliding together on the floor. When they manage to compose themselves they look at the depth gauge. It reads 1,575 feet. They are trapped, at a depth twice as deep as any previous submarine rescue. If the water above were a building, it would be taller by an extra ten floors than the Empire State Building. For ten seconds both men are silent as they wait to see if *Pisces III* will now tip forwards or back. The sub remains fixed.

In an act of controlled diligence, Chapman, who is only just keeping a lid on the terror he feels deep inside, reaches for his notebook and black felt-tip marker and writes, 'On Bottom.'

PART I

THE CABLE AND THE SUBMARINE

CHAPTER ONE

The hook resembled an anchor but with five arms instead of only two, the hands at the end of the arms resembling shovels in shape. To the hook was connected a thick wire rope, more than five miles long. Hurled overboard from the deck of the *Great Eastern*, the world's largest steamship – designed by Isambard Kingdom Brunel – the hook would take two hours to reach the floor of the Atlantic, the crew smoking clay tobacco pipes as they waited for the rope racing overboard to come to rest. When the rope was still and steady, the *Great Eastern* would move northwards, at a right angle to the line of throw, dragging the giant iron anchor along the ocean floor in pursuit of the lost transatlantic telegram cable. The magnitude of the task of fishing for the cable, snapped and lost in the ocean depths, would be picturesquely described 'as if an alpine hunter stood on the summit of Mont Blanc and cast a line into the Vale of Chamouni'.

The following morning, after 12 hours' sailing, the wire rope quivered. The anchor was caught on something and when the crew began to raise it, the object grew heavier the higher it was raised to the surface. It was the cable increasing in weight as more of its length was raised off the seabed. The wire rope had been manufactured in lengths of 100 fathoms, joined by a metal shackle. When the first shackle broke on board, the brakeman controlling the winch was quick enough to clamp it down before it raced freely overboard. The broken end whipped at the hands and faces of two crew, leaving bloody, lacerating wounds. As the cable rose three-quarters of a mile off the seabed, a second shackle broke too quickly to be caught and took two miles of wire rope back to the bottom.

The cable was hooked a second time, then lost at a raised depth of one mile. A third attempt failed and a fourth saw a shackle snap as the cable passed 800 fathoms. As the length of remaining rope was now too short to reach the bottom, the rescue operation was called off, postponed till the following summer of 1867. As the *Great Eastern* prepared to depart the mid-Atlantic for Britain, Cyrus Field, a quiet, thoughtful, smartly dressed man in his early 50s, came on deck to look overboard at the milky blue waters and to reflect on how far he had come and how far there still was to go.

Over a century before the two-man crew of *Pisces III* were charged with burying the CANTAT-2 cable, a modern successor to the lost telegram cable, the concept of transatlantic communication was but a distant dream. The decision to cast 'a thread across the ocean', as the original transatlantic cable would be called, was made in the library of a grand house facing Gramercy Park in Manhattan. The owner was Cyrus Field, a man of taste, vision and, surprisingly among the entrepreneurs and tycoons of the golden age of American capitalism, fiduciary honour. The financial crash of 1837, which wiped out 70 per cent of all businesses in New York, had presented him with an opportunity to acquire a paper wholesale firm, E. Root & Co., where he had started his career as a junior partner, by paying 30 cents on the dollar for its debts. Fifteen years later, during which time he had built the new entity Cyrus W. Field & Co. into America's largest paper and printing supply firm, he tracked down everyone to whom he had paid 30 per cent of their original debt and reimbursed them in full.

Born on 30 November 1819 into a prominent New England family, Cyrus was the son of the preacher David Field, whose family could be traced to astronomer John Field, who introduced the theories of Copernicus into England in the 16th century. Cyrus's father was fond of quoting a line from *The Pilgrim's Progress* by John Bunyan: 'To know is a thing which pleaseth talkers and boasters: to do is that which pleaseth God.' Cyrus was raised to be a doer. He began an apprenticeship at A.T. Stewart, Manhattan's leading dry goods store, at the age of 16, took night classes in book-keeping and, out on the road, proved to be a successful sales-man of pith and charm, on whose regular route he met his wife-to-be Mary Stone.

Within a decade of purchasing E. Root & Co., Cyrus Field had become one of the richest men in New York. He was also bored, so he handed over management of the firm to his brother David and embarked on a grand tour of South America to collect specimens of the continent's fauna. Upon his return in October 1853, his luggage included 20 parrots and parakeets, a jaguar that had to be restrained on a strong leash and a 14-year-old boy called Marcos – the son of one of his guides, a bullfighter in Colombia – who moved in with the family to be educated in America. (The relationship did not take. Field's daughter noted: 'A civilised life was not attractive to him,' and when her father was absent on business, Marcos was quietly shipped back home.)

Field was, however, ready for a new challenge. The opportunity to change the history of global communications, a rather leftfield idea for a retail entrepreneur, came when his brother Matthew, who had left the family business to train as an engineer, introduced him to Frederick Gisborne, a Canadian engineer Matthew had met in the lobby of the Astor Hotel on Broadway. Gisborne was in desperate need of investors with deep pockets. The former chief operator for the Montreal Telegraph Company and later Nova Scotia Telegraph Company, Gisborne, who was born in England, had embarked on an ambitious and arduous new business venture to run a telegraph cable from St John's in Newfoundland to Cape Ray on the island's most westerly tip.

As Gisborne explained to Cyrus over brandies in front of the fire in the library of the brownstone mansion in Gramercy Park, initial construction had been beset with difficulty. As the soil was only inches deep across Newfoundland, the telegraph poles had to be erected then supported by extensive rock piles. The initial plan to bury the cable was abandoned on account of the rocky terrain. Delays led to spiralling costs and, eventually, bankruptcy, the seizure of his assets and arrest. Gisborne had, prior to his detention, attempted to pay all his workers, which earned him Field's respect if not his support. Gisborne insisted that running a telegraph cable from St John's, Newfoundland's capital and the first port in

North America, to Cape Ray would speed up communications between Europe and the New World by 24 hours. Field thanked Gisborne for his time and showed him to the door, warmed by brandy but without the cold cash he so desperately required.

Returning to the library, Field lingered by his large globe, which seemed to pulse with promise. Spinning the globe around to Newfoundland, Field traced his finger from St John's, not to New York, but back out across the Atlantic to the west coast of Ireland. Why run a telegraph cable that can speed up communications by hours when you could lay one that would speed up the news by five days? London was the centre of the financial world. To control instant communication with the city's financial markets would be an exceedingly profitable prize. A letter written the following day to Samuel Morse, who had patented his telegraphy system, brought swift confirmation by return post of the feasibility of Field's ambition. Thirteen years earlier in 1843, Morse had written to the Secretary of the Treasury John C. Spencer to state that:

> a telegraphic communication on the electro-magnetic plan may with certainty be established across the Atlantic. Startling as this may now seem, I am confident that the time will come when this project will be realised.

Field then wrote a second letter, to Lieutenant Matthew Fontaine Maury of the United States Navy. Rendered lame in a stagecoach accident in 1839, Maury had embraced a deskbound life as head of the Depot of Charts and Instruments, where he devoted himself to the scientific study and documentation of the winds and currents of the world's oceans. The author of *The Physical Geography of the Sea*, a 19th-century bestseller, Maury sent Field a paper he had recently written for the secretary of the Navy. The US naval ship the *Dolphin* had already taken deep-sea soundings between Newfoundland and Ireland.

As Maury wrote:

This line of deep-sea soundings seems to be decisive of the question of the practicability of a submarine telegraph between the two continents, in so far as the bottom of the deep sea is concerned. From Newfoundland to Ireland, the distance between the nearest points is about sixteen hundred miles; and the bottom of the sea between the two places is a plateau, which seems to have been placed there especially for the purpose of holding the wires of a submarine telegraph, and of keeping them out of harm's way. It is neither too deep, nor too shallow; yet it is so deep that the wires but once landed, will remain forever beyond the reach of vessel's anchors, icebergs, and disturbances of any kind, and so shallow, that the wires may be readily lodged upon the bottom.

The depth of this plateau is quite regular, gradually increasing from the shores of Newfoundland to the depth of from fifteen hundred to two thousand fathoms, as you approach the other side. I [do not] pretend to consider the question as to the possibility of finding a time calm enough, the sea smooth enough, a wire long enough, a ship big enough, to lay a coil of wire sixteen hundred miles in length; though I have no fear but that the enterprise and

ingenuity of the age, whenever called on with these problems, will be ready with a satisfactory solution.

Samples revealed the seabed on the plateau consisted not of
sand or gravel but microscopic shells from the remains of
plankton that tumbled into the depths after they died. There
was also no evidence of strong ocean currents liable to shift
the cable once in place.

Having resolved the feasibility of the project to his satisfac-
tion, Field's next consideration was how to fund it. He began
with his neighbours. In the affluent neighbourhood of
Gramercy Park in the 1850s, one could borrow tens of thousands of dollars as easily as others in a less salubrious locale
cadge a cup of sugar. Among his neighbours was Peter
Cooper, inventor of the world's first washing machine, the
first locomotive built in the US and one of the nation's richest
men. Cooper recognised the risk – but also the opportunity
– and agreed, on condition that others help bear the financial
load. Among the early investors depicted in *The Atlantic
Cable Projectors*, an oil painting by Daniel Huntington, was
Cooper, alongside Moses Taylor, bank president and investor,
and Chandler White, a fellow paper magnet.

The starting capital now accrued, Field's first step was for
the investors to take over Frederic Gisborne's company, the
Newfoundland Electric Telegraph Company, and its debts of
$50,000, from which a new corporate entity emerged: the
New York, Newfoundland, and London Telegraph Company.
The government of Newfoundland was persuaded by Field,
who had lost none of his salesmanship, to grant the new
company monopoly rights for 50 years to lay telegraph lines

on the island. The estimated construction costs of the transatlantic cable was $1.5 million, the equivalent of 2 per cent of the entire expenditure of the US federal government. As Field later wrote, 'God knows that none of us were aware of what we had undertaken to accomplish.'

The venture was to be broken into stages, the first of which was to link New York to Newfoundland by the summer of 1854, according to the investors' prospectus, although the island's geology and geography were stonily indifferent to deadlines. The work proved interminable, beset by deep snow in winter and heavy summer rains, and it took over a year at a cost of $500,000, a third of their estimated total budget. By March 1855, when Field wrote to his brother Henry, who was leading the project, to enquire how many months more, the curt reply came: 'Let's say how many years!'

The cable to cross the Cabot Strait and so connect Newfoundland to Nova Scotia was to be manufactured in England, which had pioneered 'economic botany' and held a monopoly on gutta-percha, made from the milky sap of the gutta-percha trees native to Malaya. The sap becomes soft and highly pliable when heated above 100°F (37°C), but at room temperature it's hard and solid, maintaining enough flexibility to make it the perfect waterproof coating and electrical insulator for sheathing conductive copper cables. Although the German Werner von Siemens had developed a press to sheathe gutta-percha around a copper cable, it was the Gutta Percha Company of London that secured a near monopoly on the importation of the sap and to whom Field now turned.

In England, Field met with John Brett, founder of the Magnetic Telegraph Company, and veteran of underwater telegraphy, who agreed both to invest in the venture and advise on the construction of the cable. This was to consist of three copper wires, each individually wrapped in a sheath of gutta-percha, then collectively lashed together using tarred hemp around which another layer of gutta-percha was wrapped, with the whole package sheathed again in galvanised iron wire. Kuper and Company were contracted to make the cable and a 500-ton brig, the *Sarah L. Bryant*, hired to lay it.

By March 1855 Field had returned to New York, and with the delay on the land section from St John's to Cape Ray, it was decided to push on with laying the underwater cable over the 85-mile sound between Newfoundland and Nova Scotia. A coastal steamer, the *James Adger*, was rented to fulfil a dual task: both tow the *Sarah L. Bryant* as it laid the cable, while also providing a more luxurious accommodation for the directors, their wives, children and invited guests, who envisioned a sun-dappled holiday of chilled champagne and parasols on deck as their future fortune spooled out behind them.

This, unfortunately, was to fail to consider the contribution of Captain Turner, master of the *James Adger* and a man so intransigent to instruction as to deliberately sabotage the task at hand. Told by Field to maintain a tight course on the shortest point, he overshot on both sides of the instructed line, with the result that when just nine miles from shore they had already laid over 25 miles of cable. Then the weather turned foul, prompting Turner to abandon the *Sarah L. Bryant* in heavy fog, and when a violent gale blew up the cable was cut after it threatened to drag the ship under. (Turner turned out to be an inveterate trouble maker, later being one of the rebels who fired on Fort Sumter, and so help trigger the American Civil War.)

The cable laying, or sinking – to be more accurate – cost the company $351,000 (the equivalent of $10.5 million in 2020). More money was required and so Field set sail once again for England to raise capital in London's financial markets. In 1844 Parliament had passed the general incorporation law, which led to a substantial rise in both the number of companies and investors, yet Field's principal goal was to focus on one investor and then let the others slot in behind. His first meeting was with the Foreign Secretary George Villiers, the Earl of Clarendon, for he sought the British government as a silent and supportive partner. Britain had form in this field: together with France they had laid a cable across the Black Sea, at that time the longest underwater cable in the world. The meeting was a success, with the British government pledging both a naval ship to carry out further soundings on the seabed and an annual payment of £14,000 to use the cable. The British did insist on a provision that government messages take priority over all others, except those of the US government if they agreed to a similar contract. (They did. Eventually.)

Buoyed by the support of the British government, Field set out on a national tour to promote investment in the Atlantic Telegraph Company, which he had incorporated in London in October 1856. The project triggered both delight and disdain. The delight came from the public, who were enchanted by the idea, writing to newspapers with suggestions that the cable be suspended in the air by a phalanx of hot-air balloons, while others imagined it bobbing on the surface connected to a string of buoys at which passing ships could stop to cable messages home. Prince Albert went so far as to suggest that the cable should be encased in a glass tube. The disdain and dismissal issued from the very top, as the Astronomer Royal Sir George Airy argued that it was 'a mathematical impossibility to submerge a cable successfully at so great a depth and if it were possible, no signals could be transmitted through so great a depth'. Yet, as Field explained during his presentations in London, Liverpool, Manchester and Glasgow, Samuel Morse had already calculated that 200 signals a minute could be transmitted along the 2,000-mile wire. The stock issue was an immediate success, with 350 shares sold at £1,000 each and such public figures as the author William Makepeace Thackeray among the investors.

In the spring of 1857 the Royal Navy sent HMS *Cyclops* to survey the route, and the findings confirmed those of their American counterparts – that between Ireland and Newfoundland lay an uninterrupted underwater plateau of level ground, the exception to which was 200 miles off the coast of Ireland, where in just 12 miles the ground fell from 550 fathoms to 1,750 fathoms. The US Navy had obtained samples from the seabed at a depth of two and half miles showing that the route was covered by an ooze of decomposed plankton.

* * *

In the States, 'cable fever' was on the rise among the public, while the nation's press bestowed on the cable the reverential qualities of a Second Coming. What Christ failed to achieve on his first appearance would be achieved by a telegraph cable uniting the Old World and the New. The *New York Evening Post* declared, 'The great heart of humanity will beat with a single pulse'; The *New York Herald* called the cable, 'The grandest work which has ever been attempted by the genius and enterprise of man'; and a magazine editor in Tennessee prophesied an end to conflict: 'Wars are to cease. The kingdom of peace will be set up.'

For continent to simply speak to continent, much less communicate an overarching framework for world peace, messages would have to pass clearly along a cable 2,000 miles long, and while Samuel Morse had successfully transmitted messages through a series of ten 200-mile-long cables, this had been achieved on land. Air, unlike sea water, is not a good conductor of electricity, a factor of considerable consequence for the plan.

In a vacuum, electricity will flow through a wire at close to the speed of light, 186,000 miles per second. When electricity flows through a wire exposed to air the speed drops to between 1 and 10 per cent of the speed of light, between 1,800 miles per second and 18,000 miles per second, the exact speed being dependent upon the properties, thickness and capacity of the wire. A longer wire requires more electricity to push it along and fill it, much as a water pipe must be filled to capacity with water before any will emerge from its end. In the early days of telegraphy wires were so short that they would operate at maximum speed almost instantaneously. However, a telegraph wire in sea water will also see its capacity increase by a factor of twenty, which means the signal will slow down in what is known as 'retardation'.

The debate over the size of the transatlantic cable was a game of doubles. On one side was Samuel Morse and Dr Edward Whitehouse, hired as chief electrician for the Atlantic Telegraph Company, who believed that the wire should be as short as possible. Both men lent for support on the words of Michael Faraday who had stated, 'The larger the wire, the more electricity was required to charge it; and the greater the retardation of that electric impulse which should be occupied in sending the charge forward.'

On the opposing side was a dynamic team that mixed youth with experience. Charles Bright was only 24 years old when appointed chief engineer for the project. His early success was swift and illuminating. At 19 he had succeeded in wiring up a complete system of telegraph wires under Manchester's streets in a single night, and since joining the Atlantic Telegraph Company had chosen the landing spot in Ireland for the cable, Valentia Bay in County Kerry, with its smooth beach and adequate protection against the heavy Atlantic breakers. He was supported by the newest member of the company's board, William Thomson, chair of natural philosophy at Glasgow University and founder of the world's first physics laboratory in which students conducted physical experiments instead of only poring over textbooks. Six years earlier, Thomson had presented a paper to the Royal Society of Edinburgh in which he was the first to state the Second Law of Thermodynamics.

Bright and Thomson favoured a cable with a large diameter core to reduce resistance and the purest copper to lessen retardation. Their favoured design of cable would weigh 392 lbs per nautical mile, with the same weight again of gutta-percha insulation.

Morse and Whitehouse were in favour of a design that was a quarter the size and weight. The cable, already in production, was manufactured from seven strands of copper wire, each 0.28 inches in diameter, then twisted into a wire 0.83 inches wide, which was wrapped in three separate layers of gutta-percha, then covered in hemp and treated with a mixture of pitch, wax, linseed oil and tar. The final weight was 107 lbs per nautical mile. As this was the design that was already being made, Field and his fellow directors were unwilling to cover the costs of a redesign. It was as if a game of doubles had been played and won an hour before Bright and Thomson arrived on the tennis court.

When completed, the transatlantic cable was 2,500 miles long and weighed one ton per mile. As no vessel then afloat could carry a weight of 2,500 tons, the cable was split between the USS *Niagara*, the largest warship in the world, and HMS *Agamemnon*, an old-fashioned sailing ship that would not have looked unfamiliar to veterans of the Napoleonic Wars. The *Niagara* proved too large to dock in Greenwich, where she was due to collect 1,250 tons of cable, so the loading was switched to Birkenhead near Liverpool, with *Agamemnon* sailing into Greenwich. Once both ships were loaded they sailed to the rendezvous point, a quarter of a mile off Queenstown on Ireland's southern coast – the country's principal transatlantic port – with the sailors on each vessel offering up three cheers when the vessels finally met.

The end of the cable from *Agamemnon* was then rowed across to the *Niagara* and temporarily spliced together. A transmitter was fixed to one end and a galvanometer, designed to measure electric current, attached to the other end. There was silence on board both ships as the transmitter was switched on, then cheers of delight and celebration when the galvanometer recorded that the current had successfully passed along all 2,500 miles of cable. On 5 August 1857 the first ten miles of heavily armoured cable, for use on land and weighting nine tons per mile, was run through the Irish countryside, watched by a large crowd and a local priest intoning prayers. Cyrus Field seemed to have the last word, telling the crowd: 'What God has joined together, let no man put asunder.'

God, however, had other plans. On the first day HMS *Agamemnon* sailed five miles out, only for the heavy shore cable to become twisted in among the machinery then break, forcing a return to Valentia to begin again. By 8 August they had laid 85 miles of cable in water that was becoming progressively deeper, increasing the strain on the machinery. The deeper the cable had to fall through the fathoms, the greater its aggregate weight before coming to rest on the seabed. This increased weight meant the brakes on the mechanical device paying out the cable had to increase in pressure to maintain a consistent speed, lest gravity take hold. The brakes began applying 700 psi, rising to 1,500 psi, and by the time they were 214 miles from shore, in depths of

2,000 fathoms, the brake pressure was at 2,000 psi. Between 9 and 10 August the pressure steadily climbed to 3,500 psi, and then, just before 4 am, the cable snapped with a sound like a rifle shot.

The vessel had been in a sea trough, and when the ship rose up the tension on the cable, like a tug on a fisherman's line, triggered the snap. Four hundred miles of cable were lost. There was no choice but to postpone the next attempt until the following year. The public was understanding. A popular nursery rhyme was sympathetically adapted to recognise their plight:

Pay it out! Oh, pay it out!
As long as you are able
For if you put the damned brake on,
Pop goes the cable!

The autumn of 1857 saw the global economy slump, but Field's reputation as a man who always paid his dues ensured the company was able to remain afloat while new staff came on board. William E. Everett secured leave from the US Navy to work on the project full-time; named chief engineer, he moved to London and spent six months redesigning the paying-out machinery and brake system, resulting in a lighter, smaller and more efficient model. Meanwhile William Thomson, who reluctantly accepted that the loss of 400 miles of cable was not enough to initiate a complete rethink and a replacement with his own design, sought to examine the conductivity of the remaining cable, then cut out and replace sections that performed poorly.

The second attempt at laying the line began in the spring of 1858, when the *Niagara* and *Agamemnon* arrived in Plymouth to take on the cable. A crew of 45 men working day and night at a rate of 30 miles per day took six weeks to load the cable into the holds of both ships. For the new attempt, Field decided not to begin in Ireland and head west, but instead to lay both halves at the same time, a strategy previously promoted by the engineers that had been over-ruled by the electricians. On 10 June both ships and accompanying vessels set sail for the mid-point in the Atlantic: 52°2'N, 38°18'W. 'Never did a voyage begin with better omens,' jotted Henry Field, Cyrus's brother, about the fair winds that accompanied their journey's start. They then sailed into one of the worst storms in Atlantic history.

Agamemnon had 250 tons of cable on the forward deck, piled high above the ship's centre of gravity, and at the storm's peak the ship rolled 30° on either side, with the masts bending, almost breaking in the wind. 'The *Agamemnon* took to violent pitching, plunging so steadily into the trough of the sea as if she meant to break her back and lay the Atlantic cable in a heap,' wrote one witness. The storm lasted six days and when Field, on board the *Niagara*, whose modern design proved better suited to the rough seas, rowed across to the *Agamemnon*, he commented that it was as if the vessel had been in a pitched sea battle, with scores of men enduring injuries of varying degrees of severity.

On 25 June, 16 days after leaving Portsmouth, the ships were positioned stern to stern and, once again, the cable from *Niagara* was brought aboard *Agamemnon* and the two ends spliced together. This time a sixpence was fitted into the splice device to bring – or buy – good luck. Either way, it would soon prove an investment of limited value. Both vessels set off in opposite directions spooling out the cable, which was monitored to ensure a steady signal, but after forty miles the signal went dead. The ships returned to the original rendezvous point and prepared to begin again. It was agreed that if the signal died or the cable broke within the first 100 miles they would return to the same spot and start yet again, but if the breakage occurred beyond this point they would abandon the attempt and return to Ireland. The cable did break, but just outside the 100 mile mark and, as previously decided, the captain of the *Niagara* sailed on to Queenstown, while the captain of the *Agamemnon* returned to the mid-point and spent eight days waiting in vain for his sister ship. The second attempt had ended in failure.

For the transatlantic cable, third time was the charm. On 17 July 1859 the ships set off from the Cove of Cork, but this time there were no gathered crowds or rousing speeches. The public had lost interest and so had the press. After rendez-vousing at their designated mid-point in the North Atlantic the cable was once more spliced together, and both ships set sail, each to their own distant shore: the *Niagara* to Trinity Bay in Newfoundland and the *Agamemnon* to Valentia in Ireland. The crossings were steady, uneventful and successful. The *Niagara* arrived at 1.45 am on 5 August; when the *Agamemnon* arrived, the crowds that had been conspicuous by their absence three weeks earlier were now so great that a fight broke out between those eager for the historical honour of carrying the cable ashore and placing it in the designated trench. Charles Bright telegraphed the board of directors:

The *Agamemnon* has arrived in Valentia, and we are about to land the cable. The *Niagara* is in Trinity Bay, Newfoundland. There are good signals between the ships.

News that the cable had been successfully laid triggered a competition among newspapers around the world for the most grandiloquent leader columns. The *New York Herald* described the cable as the Angel in the Book of Revelation, with one foot on sea and one foot on land, proclaiming that Time is no longer. A wonderful cartoon widely published showed Neptune bursting through the surface, having been electrocuted by the cable and clutching a fistful of telegram messages. *The Times* of London declared: 'The Atlantic is dried up, and we become in reality as well as in wish one country.'

Although the current was flowing along the cable, no message had yet been sent. Field informed the press that the telegraph instruments were in the process of careful adjustment and that the first message would be from Queen Victoria to President Buchanan. On 16 August the message was sent. And received.

To the President of the United States, Washington:
The Queen desires to congratulate the President upon the successful completion of this great international work, in which the Queen has taken the deepest interest. The Queen is convinced that the President will join with her in fervently hoping that the electric cable will prove an additional link between the nations, whose friendship is founded upon their common interest and reciprocal esteem.

A 100-gun salute took place at New York's City Hall when confirmation of Victoria's message was received, while the remaining parts of the cable were sold to Tiffany & Co., who proceeded to cut them into short lengths bound in brass and sold them on as a profitable souvenir.

What the public were not told was that the royal message took almost 17 hours to transmit and be received, a country mile from the 200 messages a minute promised by Morse. Over the next few days messages came through, including news of a small collision involving a Cunard liner, but each message was weaker and more unintelligible than the last.

By 1 September the transatlantic cable was dead. The report into the failure had more words than the Bible but its conclusion was straightforward: the cable had most likely burned out by a sustained voltage greater than its capability.

The American Civil War would begin and end before Cyrus Field again attempted to lay a cable across the Atlantic. The construction of the *Great Eastern*, the largest steamship afloat, meant that a single vessel could carry all 2,700 miles of cable within a hold the height of a cathedral. This time William Thomson and Charles Bright designed the cable, which weighed three times as much as the original, and consisted of seven strands of copper wire and four layers of gutta-perch rather than three. William Howard Russell of *The Times*, who made his name and largely invented the term 'war correspondent' while covering the Crimean War, was invited to accompany Brunel's leviathan, which set off from Foilhummerum Bay on Valencia Island for Heart's Content in Newfoundland on 23 July 1865. As the captain recorded: 'I had no mind, no soul, no sleep, that was not tinged with the cable.'

Field spent much of the trip literally in the dark. In the testing room, behind blackout curtains, he stared obsessively at the light of the mirror galvanometer, designed by Thomson to more accurately detect a weakening signal. A gong was struck to alert the captain and so stop the ship when a fault was discovered. The first gong was struck after 84 miles when the signal weakened. Ten miles of cable were hauled back up and re-examined till the culprit was uncovered – a small piece of wire penetrating into the core. As Russell reported, this was 'flagrant evidence of mischief'. In fact, after the same thing happened on two further occasions, the culprit was identified as a fixable fault in the device paying out the cable.

On 2 August, when the *Great Eastern* was 600 miles from Newfoundland, the cable snapped. For eleven days the crew made repeated attempts to find, hook, then raise the cable, before finally running out of rope. On the journey home Field was cocooned in a cloud of depression, dispelled by the warmth of affection that greeted the ship on arrival. As there had been no news for over two weeks, the *Great Eastern* was thought sunk with the loss of all on board. Relief at their survival was therefore the principal emotion, not anger at another failed attempt. In fact, such was the goodwill that Field was able to swiftly raise funds to not only raise the lost cable but to also lay a 'second' cable.

In the end, the second cable was laid before an attempt to recover the first. Laid down on a course 30 miles south of the first cable, the second cable was a smooth, swift success. On 21 July 1866 the *Great Eastern* passed the midway point, and by the time the vessel approached the Newfoundland Field, the ship was able to get messages on the latest news and stock-market prices. By 2 August the transatlantic cable, after a seven-year absence, was back in business. Yet Field had no time to savour his success, as he wrote to his wife: 'We leave in about a week to recover the cable of last year.'

Raising the first lost transatlantic cable was a painful lesson in perseverance. Two sister ships, the *Albany* and the *Terrible*, found the cable on around 10 August 1866, managed to raise it two hundred fathoms, and when bad weather broke they secured it with buoys. The weather then broke off the buoys, sinking both cable and the wire rope to which it was attached. The *Great Eastern* arrived on 12 August and joined what became a four-day search. On 20 August the crew snared the cable and spent 14 hours raising it to the surface. As Field wrote:

> We had it in full sight for five minutes. A long, slimy monster, fresh from the ooze of the ocean's bed – but our men began to cheer so wildly that it seemed to be frightened, and suddenly broke away and went down into the sea.

Over the next few days the cable was hooked a number of times but then broke under its own weight before reaching the surface. A new approach was devised. The ships moved to slightly shallower waters. The *Great Eastern* hooked the cable, raised it to just 900 fathoms beneath the surface, supported it with buoys, then sailed on a few miles west and raised another portion. A fourth ship in the fleet, the *Medway*, two miles away, raised a different portion of the cable, then deliberately broke it to lighten the weight on the section under the control of the *Great Eastern*.

At 1 am on Sunday 2 September the cable was finally hauled aboard the *Great Eastern*. Once fixed up to the testing room, a message was sent to Ireland. A reporter from *The Spectator* described how for over a year the cable was tested day and night, revealing 'wild incoherent messages from the deep ... merely the result of magnetic storms and earth-currents', but now what emerged was 'the first rational words uttered by a high-fevered patient, when the ravings have ceased and his consciousness returns'. The success was reported to the other ships in the convoy by the release of a barrage of rockets.

Watching the night sky illuminated by white light and the jubilation of the crew, Field was moved to seek solitude:

> I left the room, I went to my cabin, I locked the door; I could no longer restrain my tears – crying like a child, and full of gratitude to God that I had been permitted to witness the recovery of the cable.

The recovery of the cable meant that soon there would be not one working transatlantic cable but two. A revolution in global communications had been achieved.

CHAPTER TWO

The first British casualty to the submarine life was John Day, an illiterate labourer to a ship's carpenter, who lived in Norfolk in the 18th century, bet his life on his revolutionary diving design and lost. There were earlier pioneers who made elaborate claims of aquatic submersion but the suspicion of doubt over their exploits lingers for the simple fact that they survived.

Let us take, for example, the Dutchman Cornelis Drebbel, who in the 1620s was said to have created an early submersible that was essentially one wooden rowing boat fixed on top of another, covered in leather skins, with six paddle holes punched through the wood, then fixed with a watertight leather seal. According to reports, he and twelve oarsman sank to a depth of 15 feet and then, thus submerged, travelled the Thames from Westminster to Greenwich. So intrigued was King James I (and VI of Scotland) that it was claimed he accompanied Drebbel on a dive in the Thames. Unfortunately, today's submarine historians are sceptical of Drebbel's claims to the point of abject dismissal, believing at best the vessel only partly submerged. Drebbel lived to tell his tale, tall or not, and died as a pub landlord in 1634.

The lure of the deep has long cast a spell over men. Alexander the Great was said to have been lowered in a primitive diving bell from a galley anchored in the port of Tyre. Leonardo da Vinci made a series of sketches for a submersible, fitted with fins on top and bottom and armed with an augur designed to drill holes in the hulls of enemy vessels. The designs were for the Doge of Venice, but so perturbed was Leonardo by the lethal potential of his imagination that he kept them secret 'on account of the evil nature of men, who would practise assassinations at the bottom of the seas by breaking ships in their lowest parts and sinking them with the crews who are in them'. Leonardo also quickly grasped the concepts of pressure and the weight of water:

> If all the bed of the sea were covered with men lying down these men would sustain the whole of the element of water, consequently each man would find that he had a column of water a mile long on his back.

If the submarine has a father it could be argued that it was Archimedes, who while wallowing in the bath discovered the principle of buoyancy by displacing water equivalent to his own weight in the original 'Eureka!' moment. Today the principle is defined as 'any object, wholly or partially immersed in a fluid, is buoyed up by a force equal to the weight of the fluid displaced by the object'.

Over the centuries, achieving three states of buoyancy was crucial to the success of a submarine: positive buoyancy, where the density of the vessel is less than the surrounding water and it will thus float on the surface; negative buoyancy, where the vessel's density is greater and so it will sink; and the elusive neutral buoyancy, where the density of the craft is now equal to the surrounding water and, although submerged, it is capable of remaining in a fixed position.

In 1578 William Bourne, an English mathematician and retired naval gunner, published plans for a wooden submersible to be powered by oarsman and able to alter its own buoyancy via a screw-operated ballast system. Wooden tanks fixed to the side enabled water to pour in and weigh it down, which could then be forced out, allowing air in and the vessel to rise to the surface. In the design, which was never built, a hollow mast would act as a snorkel.

Two centuries later, in 1774, John Day would succumb to the pressures of the deep. A solitary man who enjoyed the darkness – he was once rescued from a potholing misadventure in Peak Cavern in Derbyshire – Day was inspired in his aquatic scheme when he realised that a wooden barrel, appropriately sealed, would keep water out but also air in. He subsequently adapted the idea to a large wooden box weighed down with stones that could then be released to assist in surfacing. The box was fitted onto a small market boat and Day sank to a depth of 30 feet at the Yarmouth Broads, then surfaced to the delight of the gathered crowd.

After Day's first successful submergence, he persuaded the noted gambler Christopher Blake to advance him £350 to build a larger model with the view to allowing the public to bet on his chances of surviving 24 hours at the bottom of the sea. 'I found an affair by which thousands may be won; it is a very paradoxical nature that can be performed with ease,' explained Day to Blake in a letter in which he asked for £100 of every £1,000 Blake won on the bet.

Blake agreed to back Day, but advised that he limit the submergence to 12 hours at a maximum depth of 120 feet and 'at any expense to fortify the chamber'. The funds enabled Day to purchase the *Maria*, a 50-foot sloop that he fitted with a watertight cabin. Iron buckets were fitted to either side of the keel and 20 tons of rocks added that could be later released by Day when the time came to ascend. The cabin was 12 feet long, nine feet broad and eight feet deep, and by Day's calculation could hold '75 hogshead of air'. Positioned two feet above the main deck, the contraption was to be entered by Day via a bevelled hatch. A hammock was available on which he could while away the hours sipping from a flask of water and munching on a day's ration of ship's biscuits. Communication with the surface was via the release of three coloured buoys: the white one signalled 'All is well'; red 'Indifferent health'; and black 'In great danger'.

At 2 pm on 28 June 1774 the *Maria* was towed out to a spot just off Drake's Island in Plymouth Sound by HMS *Orpheus*, a naval frigate on whose stern gathered invited guests. Hundreds more members of the public were spread out on the coastline in the hope of witnessing Day sink, then ascend at the appointed hour. A contemporary account described Day as 'more than ordinarily cheerful' and 'confident that his enterprise would be crowned with success and universal acclamation'. In the cabin Day had set up a clock to alert him to the correct time to turn the iron rods, release the ballast, and ascend to celebrity and a fortune in wagers.

At first the *Maria* failed to sink when released. After an extra load of 20 tons of rocks was added, she disappeared below the waters. Sadly, the *Maria* never resurfaced, the buoys never bobbed up and Day was never seen again. Instead, eyewitnesses on the bow of the *Orpheus* said 'a number of very large bubbles kept rising from the bottom, and the sea became covered with white froth for some yards around'. The weight of water at a depth of 132 feet is believed to have either crushed the cabin or splintered it enough to let in water. Nikolai Falck, a Dutch doctor who claimed to have the ability to resurrect Day if the vessel and body could be retrieved, instead recorded that Day had 'descended ... into perpetual night'.

Across the Atlantic, conflict prompted the submarine's progress. While a student at Yale College, David Bushnell used his extracurricular studies to prove that the explosive effect of gunpowder while submerged in water was much greater than on dry land. Later, during the War of Independence, Bushnell invented *Turtle*, a one-manned submarine devised to creep up on Royal Navy ships, burrow through their hull with an augur and pack the hole with explosives on a timed charge, allowing the sub to silently slip away to a safe distance before any subsequent explosion. This was the plan on paper, but in practice it would be a different matter.

Turtle comprised two large pieces of oak, hollowed out and then sealed together using a thick iron band, with access through a hatch on the top. Inside were hand-cranked propellers: the one on the side provided horizontal motion, the one on the top vertical motion. There was a single rudder for steering and the pilot was equipped with a small viewing port. Buoyancy was controlled via a tank so that *Turtle* sank when flooded by sea water, which could then be expelled via a hand pump to permit ascent. A pipe was fitted to carry away carbon dioxide.

The preliminary designs had so dazzled General George Washington, commander in chief of the American army, that he funded the construction. *Turtle*'s first clandestine mission was to sink HMS *Eagle*, a 64-gun warship anchored in the Hudson River. The test pilot was to be Bushnell's brother Ezra, but a bout of sickness saw the substitution of Sergeant Ezra Lee into the bowels of the *Turtle*. On 6 September 1777 Lee claims to have slipped under the surface of the Hudson and come close to successfully carrying out his mission; but his attempt to drill through the hull failed as he unwittingly selected a point where it was protected either by copper sheeting or the iron plate around the ship's rudder. Unable to penetrate the hull and so attach the explosives, Lee glided silently away. Yet once more there's considerable doubt whether the story is true and if the assault was actually attempted. In his book *The Submarine Pioneers*, Richard Compton-Hall, director of the Royal Navy Submarine Museum, dismissed it as a fabricated account concocted by Lee 40 years after the event. George Washington was himself unsure of exactly what happened. Writing in 1785, he cast doubt on the exact events while still praising it as 'an effort of genius'. In a letter he insisted that Bushnell:

had a machine so contrived as to carry him underwater at any depth he chose, and for considerable time and distance, with an appendage charged with powder, which he could fasten to a ship and give fire to it in time sufficient for his returning, and by means there of destroy it, are facts.

What is indisputable is that Bushnell conceived and designed the *Turtle*. He may also have passed on some of his knowledge to Robert Fulton, a talented artist born in Pennsylvania who had considered a career as a portrait artist, before design and aquatic inventions came to dominate his working day. Fulton travelled to France, where he promised Napoleon a way to 'carry the war to the shores and ports of Great Britain'. In 1798, while living in Paris, Fulton devised both a new form of submarine and 'torpedoes' – underwater explosive projectiles – whose name he took from a genus of ray that paralyses its prey with electric shocks. The French, however, refused to fund the construction of a prototype, so Fulton raised the money himself, first by constructing a panoramic painting of Paris that he charged the public to see, then through persuading a Dutch investor to make up the shortfall.

Fulton's *Nautilus* was constructed in Rouen, measured 20 feet in length, with a six-foot beam, and was equipped with ballast tanks, a hand-cranked propeller and horizontal rudders. On the surface it was powered by sails, while for diving it had a basic periscope and snorkel. When Fulton moored *Nautilus* on the Seine in Paris, thousands flocked to see the craft, and Napoleon finally agreed to invest 10,000 francs on condition that it was based in Brest. On 3 July 1801 Fulton embarked on a test dive, sinking to a depth of 25 feet in steady increments of five feet at a time, but aware of the dangers of pressure he refused to sink any deeper. During the dive he discovered a number of design flaws, most notably that because there were no windows and therefore no light, the chamber was illuminated by candlelight, which depleted the available oxygen. Fulton subsequently fitted a window and found he was able to consult his compass from the available daylight. Assisted by a crew of three hand-cranking the propeller, he could cover a distance of 500 yards in under ten minutes.

Fulton, a suspicious man of firm intent, sensed that his dealings with the French government were becoming precarious and that the navy planned to steal his design rather than pay him to construct a fleet of similar vessels. He also met with resistance from the French minister of the Admiralty, who was concerned about the ethics and legality of submarine warfare.

So rather than waiting to be either conned or cancelled, Fulton deliberately destroyed the *Nautilus*, informing the French government that it had sprung a leak and sunk. He then left France for the Netherlands, where he was approached by secret agents offering the opportunity to design 'plunging boats' for Britain. With no allegiance other than to the furtherance of his own design, Fulton accepted and promptly took up residence in Piccadilly under the alias 'Robert Francis'. His demands were simple and direct: he wanted £7,000, access to Britain's naval dockyards and 40 tons of gunpowder. With this, Fulton insisted he would construct a new submarine vastly superior to the *Nautilus*, one capable of carrying a crew of six submerged for six weeks and travelling underwater as fast as any fishing boat on the surface.

The problem was that, like the French before them, the British soon tired of 'plunging boats'. They were, however, interested in Fulton's torpedoes. He spent the next two years refining their design before finally rendering a member of the naval high command quite literally speechless. In October 1805 the *Dorothea*, a 200-ton vessel, was anchored off the coast of Kent, the target for a test of Fulton's torpedo. A stern critic of the idea, a Captain Kingston said he had so little faith in Fulton's design that he would happily dine on board the *Dorothea*. So vast was the explosion that destroyed the ship that Kingston was rendered mute. As Fulton noted: 'Ocular demonstration is the best proof for all men.'

But just as Fulton had abandoned the French, so he did the British, leaving Europe to try his luck with his home nation: the United States. Although his agreement with the British government included a 14-year prohibition on publishing what he had gleaned while working for the Crown, Fulton published *Torpedo War and Submarine Explosions* in 1810, and by 1812, when the US was once again at war with Britain, worked with the US Navy to develop new designs, although they never left the page. Fulton died in 1815 from pneumonia, developed after rescuing a friend in difficulty in the water.

The race to conquer the depths was not restricted to Britain and the US. In 1850 Wilhelm Bauer, a Bavarian artillery corporal, built what he christened *Brandtaucher* (*Incendiary Diver*), which resembled a steel box equipped with mechanical arms capable, he hoped, of fitting explosives to the hulls of enemy vessels. *Brandtaucher* was 27 feet long, displaced 39 tons and had two small windows that, on the surface, protruded above the water line like a snout. Bauer could be hot tempered but had a cool head when required, as demonstrated when he managed to rescue himself and his two fellow occupants from 50 feet beneath Kiel harbour. *Brandtaucher* had been stuck for five hours when Bauer began to deliberately flood the sphere. His fellow submariners thought him crazy and he had to fight them off, yet he understood that it was impossible to open the hatch against the pressure of the surrounding water. Only by allowing the incoming water to compress the air inside *Brandtaucher* would it balance the pressure outside. Holding his nerve while the sphere completely flooded, he was then able to unscrew the hatch so all three could swim to the surface, making Bauer an unwitting pioneer of submarine rescue.

In 1853 Bauer moved to London. His reputation and German background led to an introduction to Prince Albert, the husband of Queen Victoria, who helped him find a position with the Scottish naval engineer John Scott Russell at the shipyard in Millwall where the *Great Eastern* was then under construction. Russell had his own ideas for a submersible vessel: a diving bell that could be 'walked' along the seabed by the leg power of its two occupants. A test dive in the Thames was only partially successful; it did sink, but then became wedged on the muddy floor, resulting in the deaths of two men.

Bauer and Russell did not get on, as Bauer believed that the Scot was stealing his ideas, so he decamped, leaving Britain for Russia, her enemy in the Crimea. Bauer based himself in St Petersburg, where he designed and built *Seeteufel* (*Sea Devil*), a vessel with a crew of 13 and ballast tanks operated by hand pumps. It was said to have made over 130 dives, though as a weapon of war it was less than effective. On a test run charged with sinking a disused vessel in Kronstadt harbour, it became stuck in the mud, and when it finally surfaced sank again when water poured into an open hatch. It was, however, the venue for the world's first underwater recital, when, to mark the coronation of his patron Tsar Alexander II, Bauer arranged for a string quartet to sink beneath the waves and serenade the Tsar from the deep.

The American Civil War was a boom period for the development of submarines. The US Navy sought to blockade the Confederate states, who, in turn, wished to both evade their enemies and, where possible, destroy them. For both these reasons the South embraced the submarine. In Richmond, the Confederate capital, a four-man submarine was successfully used in a trial to sink a large, flat-bottomed cargo ship. After approaching the ship, a diver slipped out of the sub and attached a torpedo – equipped with an electrical cable that allowed detonation at a safe distance – to the hull. A Pinkerton spy watched the trial and successful detonation, and then advised all shipping to be on the lookout for moving green buoys or floats that were, in fact, disguised air hoses.

The North had plans of their own. In March 1861 President Lincoln was approached by Brutus de Villeroi, a French citizen who had emigrated to America. De Villeroi was a professor of both drawing and mathematics, and among his students in Nantes was a young Jules Verne, who would have been aware of his tutor's inventions. In 1835 de Villeroi developed *Waterbug*, a ten-foot-long fish-shaped vessel that could carry three men and stay submerged for an hour.

By 1861 de Villeroi's plans were as expansive as his new home nation, and the design that would be dubbed *Alligator* was iron-hulled, capable of holding a crew of 12, and had a hatch that allowed a diver to slip out and access enemy shipping while the craft was still submerged. Although the US Navy were impressed, they were also in a hurry, so they gave de Villeroi 40 days to construct a new prototype adapted as a weapon of war. Unfortunately he was unable to work at speed and so was sacked from his own project; accused of being in breach of contract, he was paid nothing, and the navy brought on board Samuel Eakins, a former army ordnance expert.

By May 1862 USS *Alligator* had grown to 47 feet in length, had a beam of four feet and was propelled by hand-operated paddles by 18 oarsman, part of a crew that also included two helmsmen and two divers. The plan was to launch the sub against Confederate shipping on the James River, but this proved problematic. *Alligator* needed clear water with a depth of at least seven feet to successfully deploy, and the James River was so cluttered as to be impassable. The mission was cancelled, and over the next 12 months the submarine was adapted and refined but rarely progressed beyond sinking to the bottom of a river bed and rising again, as if propelled by the silent prayers of those on board.

In March 1863 while under tow, *Alligator* was lost in a storm off Cape Hatteras and despite attempts as recently as 2005 to find its resting place, its exact location remains a mystery. As for de Villeroi, he was disgusted by what he viewed as the treacherous and ungrateful behaviour of the Americans and returned to his mother country. He wrote to

Emperor Napoleon III with outlandish plans for a new 125-foot submarine to be armed with saws and guns, yet a naval commission charged with judging the efficacy of his ideas was damning.

The father of the modern submarine was born facing the sea. The son of a coastguard, John Philip Holland was born in 1841 and raised in the small town of Liscannor on the west coast of Ireland. On the death of his father in 1853 the family moved to Limerick, where Holland continued his education under the Christian Brothers teaching order. A brilliant scholar with an abiding interest in science and mathematics, from an early age Holland was also fascinated by the possibility of undersea travel. Although a good student, Holland made a poor teacher. When he himself joined the Christian Brothers he was unable to corral a class's attention and allowed too much misbehaviour. In 1873 he left the order and bought an Atlantic crossing.

In the United States Holland returned to the blackboard by day, teaching pupils at St. John's Parochial School in Paterson, New Jersey, but at night he worked on his submarine designs. Within two years he had constructed a 16-foot, pedal-powered submersible for which the pilot wore a diving helmet rigged to tanks of air. He approached the US Navy with high hopes, only to be dismissed with the comment, 'a fantastic scheme of a civilian landsman'.

There were, however, other more supportive if clandestine sources of investment. The American wing of the Fenians, a group dedicated to the armed struggle against the subjugation of Ireland by the British, recognised in Holland's submarine the perfect guerrilla weapon with which to attack the might of the Royal Navy. But the craft's maiden voyage on the Passaic River in May 1878 funded by the Fenians was not a success, as the holes around the screw shaft were insufficiently sealed, water poured in and Holland was mocked by spectators as the man who had 'built a coffin for himself'.

The Fenians were determined that the vessel begot coffins only for the British and were keen to put the submarine into action. Their plan was to load it onto a larger fishing vessel, anchor this inconspicuously by a British naval vessel and then release their secret weapon, which they called 'The Salt Water Enterprise'. The principal problem with their battle plan was that it did not work.

Holland persuaded them to invest in a new prototype, which he launched in 1881. This sub, which displaced 19 tons, could travel at eight knots on the surface and seven knots when submerged, and was designed to be manned by three people, whose comfort was assured by the provision of a toilet. Compressed air was used to operate the ballast tanks, and the vessel was powered by a petrol engine, the fumes of which were carried away by a flap valve. The new prototype was also equipped with a torpedo, fired using compressed air. Holland's pride in his new design clearly outweighed any need for military discretion as, following an interview he gave to the *New York Sun*, the sub was bestowed with the moniker *Fenian Ram*.

Now that their name was inextricably linked to the vessel, Holland's paymasters became increasingly possessive. Unhappy with the speed of the vessel and unwilling to tolerate any more of what they saw as the designer's prevarication and desire for further sea trials, the group stole the sub, attempted to sail it down the Hudson, terrified passers-by and accidentally grounded it. Holland, so upset by their behaviour, refused to assist in any repairs and severed their partnership.

Holland would go on to develop several new iterations of the submarine, but the design that secured him the soubriquet 'father of the modern submarine' was drawn up in 1893 for the US Navy, who had long revised their view of him as a 'civilian landsman'. The navy announced a competition and Holland, who had in the meantime set up the Holland Torpedo Boat Company, entered. On two previous occasions Holland had entered and won naval design competitions, only to see them fail to leave the page.

The Irish inventor conceived the *Holland VI*, a steam-powered submarine that displaced 64 tons on the surface, 74 tons submerged, and was capable of a maximum depth of 100 feet. The *Holland VI* was also equipped with a torpedo tube and an 18-inch torpedo. Holland's design slipped successfully off the drawing board and the vessel was launched on 17 May 1897 as the USS *Holland*. What made his design so attractive as to be duplicated by other navies around the world was a propulsion system that combined both the internal combustion engine and the electrical battery. The electric storage battery had been the invention of Gaston Planté, but it was Holland who had the vision to match it to the combustion engine as a means of successfully powering a submarine. While on the surface the engine both powered the vessel and the compressed air generator, as well as creating excess power for storage in the battery. A fully charged battery would then power the *Holland VI* for 30 miles underwater.

In the USS *Holland* the US Navy had a submarine with reliable means of propulsion, an efficient method of steering, achieved by a large rudder mounted behind the screw, and the means to both dive and surface on command. By introducing a number of ballast tanks around the vessel, the boat's buoyancy could be easily altered while its centre of gravity remained fixed. The use of a large hydroplane aft enabled the *Holland* to dive underwater like a dolphin.

So successful was Holland's submarine that the Royal Navy licensed the design and ordered five 'Holland' boats built at a cost of £175,000. The shipyard designated to construct the vessels was Vickers in Barrow-in-Furness, a seaside town in what was then Lancashire known to all and sundry as Barrow. When the submarine's new commanding officer paid his first visit to Vickers he was directed to the 'yacht shed', where the vessel, still secret, was under construction. He was initially unimpressed, describing the boat as 'like a very fat and stubby cigar'.

The first decade of the 20th century saw the submarine develop from a vessel of coastal defence to an ocean-going leviathan of 500 tons. In Britain there was resistance to their use, but a charm offensive aimed at King Edward VII by Admiral 'Jacky' Fisher, then commander-in-chief at Portsmouth, proved to be successful. The king visited one vessel and his son, the Prince of Wales and future King George V, was taken on a dive. 'Everyone was averse from the Prince's going down, but he insisted and I think he was right. It will give a lift to submarines (crafty Jacky) …' wrote the chairman of the War Office Reconstruction Committee. Fisher was not just crafty but fortunate too, the sub's crew less so, as a few days after the royal dive HMS *A1* was submerged and accidentally struck SS *Berwick Castle*, sinking the sub and killing the nine men on board.

By the outbreak of the First World War there were over 400 submarines in the world's waters, operated by 16 different navies. The most successful were Germany's *Unterseeboote* (U-boats), which could travel up to 5,000 miles at depths greater than vessels of any other nation.

John Philip Holland died on 12 August 1914 of pneumonia. He was 73 years old and had he lived just one month longer he would have witnessed the terrible debut of the age of the submarine, an age he helped create. On 5 September 1914 HMS *Pathfinder*, a 2,940-ton naval vessel, was patrolling 15 miles off St Abb's Head on the edge of the Firth of Forth. Charged with protecting the new naval dockyard at Rosyth, the elderly ship had 268 men on board, one of whom, a senior rating, spotted a periscope and then a torpedo track tearing across the grey waters 1,300 yards distant. It was already too late.

Whatever confused evasive orders were shouted, seconds later *Pathfinder* was struck a fatal blow. The explosion was so powerful that it was felt on board a fishing trawler ten miles distant. Bulkheads crumpled, the funnel tipped over and *Pathfinder* collapsed in two parts. Captain Francis Leake, the 45-year-old commanding officer, could merely shout to his crew, 'Jump, you devils, jump.' Leake would be among only 12 survivors, with 256 men under his command lost, the vessel and crew victim of the first submarine attack in naval history. The fears of Leonardo da Vinci had been brutally realised.

CHAPTER THREE

The *Pisces* submarine, which would revolutionise undersea work, was invented by three divers from Vancouver, so casual in their appearance as to be dubbed the 'T-shirt boys' and who tested their invention, unwittingly, in the exact location and time where the US Navy were deploying anti-submarine torpedoes. *Pisces* is therefore a testament both to the success of the T-shirt boys' design and the failure of the US Navy's new prototype to seek and destroy her.

Al Trice was softly spoken and in the early 1960s looked like Steve McQueen. A native of Vancouver, the pipe-smoking Trice had been introduced to scuba diving, then in its infancy, by a former British Navy frogman and was quickly hooked. At the time commercial divers dismissed the aqualung as 'mouse gear' and its aficionados as 'frogs', preferring the heavy steel helmet and fixed oxygen line of traditional 'hard-hat' diving', but Trice became a master of both systems, able to switch comfortably between the two depending on the nature and demands of the job. It was a versatility that was to make him popular with contractors and earn him a reputation as one of the most respected divers on the Pacific Coast.

Divers Don Sorte and Al Trice, the pioneers
of the *Pisces* submersibles.

Don Sorte was six foot two, with rugged good looks and an ill-fitting toupee that he liked to hurl onto the dance floor when particularly invigorated by a nightclub soundtrack. He had three poodles, each dyed a different colour – rose, blue and yellow – and each answering to 'Tiger', a novel nomenclature that he believed saved time. Call one name and all three came running. In the early days of scuba, Sorte was paid by police to retrieve accident victims from the depths but warned not to pick their pockets – they needed the wallets for identification. A move into commercial hard-hat diving was more lucrative, enabling him to eventually commission a self-portrait in which he is seen genuflecting to a giant dollar bill.

Together Trice and Sorte worked as a team on deep dives, and it was the *Barge 10* accident that convinced them of the need for a commercial submarine capable of safely working at extreme depths. In 1964 the barge, loaded with heavy bunker oil, had sunk to a depth of 330 feet, and its retrieval was then considered the deepest salvage operation in the world. Trice performed the first solo dive to survey the scene. This was to be the first of dozens of highly dangerous 'bounce' dives that involved spending as little time on the bottom as possible, followed by a long, slow and arduous ascent to the surface, then hours in a decompression chamber. At 300 feet the amount of nitrogen – from the air breathed from the aqualung – forced into the blood stream by the atmospheric pressure of water has the equivalent effect of downing four or five large Scotches, and divers could do stupid things while 'drunk' or 'narc'd' on nitrogen, including taking off their aqualung. By the end of the salvage job Trice and Sorte were spending as little as 14 minutes actually 'on site', and Trice knew there had to an easier, safer solution. Why couldn't a submarine do these jobs?

The reason, in 1964, was that commercial salvage submarines didn't exist. The nearest equivalent was the *Denise*, the small submarine used by Jacques Cousteau and launched from his support vessel *Calypso*. Cousteau and his ship had been travelling along the US West Coast that year, and the extensive TV and newspaper coverage that followed in the wake of the French undersea explorer had caught Trice's attention. Then there was the *Trieste*, designed by the Swiss scientist Auguste Piccard, which in 1960, piloted by his son Jacques and the American Don Walsh, reached the bottom of the Challenger Deep, a trench in the South Pacific over six miles deep. Yet despite its world-record-breaking achievement, the *Trieste* was unsuitable for any kind of commercial work, so Trice concluded that it would be necessary to build what he could not buy.

On a road trip inconceivable today, in our world of corporate secrecy and nefarious industrial espionage, Trice and Sorte drove down the length of California stopping off at the research laboratories of America's leading companies – Lockheed Martin, Westinghouse Electric, General Dynamics and General Mills – each of whom were developing their own portable submarine but were generous or foolhardy enough to show round what they saw as a pair of enthusiastic amateurs in sneakers and scruffy T-shirts, rather than dedicated competitors. At Lockheed's aerospace research centre the pair were even treated to a demonstration of a moon buggy seven years before David Scott and James Irwin would put one to use. Yet although inspirational, the visits revealed no key secret beyond Trice's conclusion that the final design had to be small enough to fit in the hold of a cargo plane and a small- to medium-sized support vessel.

On the road home the pair stayed with friends in Seattle, where they met the other figure in what was to become their subaquatic troika. Mack Thomson was a nudist and amateur diver, devoted to technical journals in the way other men were to *Playboy* and *Sports Illustrated*. He had made his own drysuit by stitching the plastic material harvested from baby mattresses onto a pair of thick long johns, and although he paid for a second-hand oxygen tank, he had built his own regulator. On one dive he experimented with attempting to surface at speed by using compressed air to blow up his suit, an unwise idea that introduced him to the 'bends', the painful and potentially fatal condition when bubbles of air form in the bloodstream.

Thomson had studied Auguste Piccard's designs and was aware the cabin had to be the heart around which everything else was constructed. He also had a friend, Warren Joslyn, who worked as a stress engineer at Boeing and who, after initially describing their plan as 'impossible', then set about proving himself wrong. It was Joslyn who produced the engineering specifications for a design centred around two spheres, the larger of which would house the crew and their equipment. A sphere was capable of achieving the most equitable distribution of the ocean's pressures, and Joslyn was confident his design could reach depths of 1,650 feet (500 metres). The figure laid down by the US Navy for any private commercial submarine operator who wished to bid for contracts retrieving torpedoes from test ranges was 1,970 feet (600 metres). If it could reach 500 metres, it could surely go to 600 metres.

Vancouver Iron and Engineering built the spheres for $50,000, and they were then moved to an empty back room in a mushroom cannery, with the 'rent' paid in sweat as all three helped to load crates of tinned mushrooms onto delivery trucks. When Thomson designed a part, Trice would deliberately drop it onto the cannery's concrete floor; if the part was robust enough to survive the drop they moved on to the next stage. Shortage of money forced the team to be creative. The carbon dioxide scrubbers, for instance, were powered by a small motor liberated from a Singer sewing machine that belonged to Thomson's wife.

The initial plan – to build the sub in three months on a budget of $20,000 – was impractical. Fifteen months and hundreds of thousands of dollars later, *Pisces* had yet to touch the water and when first introduced to her new environment the sub did not float level. The tail was too heavy, a problem finally solved by packing in fishing floats to increase her buoyancy. Communication was also a problem. Radio waves won't pass through water, and, while sound waves can carry for hundreds of miles, they refract and bend as they travel. The team discovered that the support ship *Hudson Explorer* had to be directly above *Pisces* to ensure comprehensible communication as the refraction was minimal on the vertical plane. Despite all these issues, *Pisces* was finally ready for an open-water trial at depth.

The decision to test their submarine in a missile range was not deliberate. The Strait of Georgia was an open body of water close enough to the port of Vancouver for ease of access, and from there the *Hudson Explorer* and *Pisces* moved to the Jervis Inlet, which offered the necessary depth. On arrival on a December morning in 1966 the team discovered a US Navy vessel was firing anti-submarine torpedoes, and although the captain was unwilling to cancel the tests to accommodate a few commercial divers and their miniature submarine, he agreed to what both teams considered a clear distance; to take a break during their descent and to alert the crew of the *Hudson Explorer* when a torpedo was in the water. To ensure safe passage Mack Thomson, the test pilot, also wore his lucky '007' James Bond sweater, although the descent would test the nerves of even the most seasoned secret agent.

Communication was patchy, but at 660 feet word came down that a torpedo had been fired and that he should switch off all equipment, so in nervous silence Thomson, who even closed his hand over his wristwatch to silence the ticking, listened as the sound of the projectile drew ever closer – then passed on by. Only then did he continue the descent on a test dive. The troubles continued: one of the two smaller 16-inch compartments used for ballast and trim control imploded due to the pressure at 1,400 feet, he struggled to release the drop weight and when he finally surfaced hours later than planned, the *Hudson Explorer* struggled to find *Pisces* in the dark. Yet despite its troubles, they were all considered fixable, and the submarine's test dive was viewed as a success.

It would not be long before Vancouver's T-shirt boys could sail into the open market as the proud inventors of the most affordable and versatile miniature submarine in the world.

* * *

When Vickers wished to establish themselves in London they built the tallest building in Britain. The Vickers Tower rose almost 400 feet above the north bank of the River Thames, between Lambeth Bridge and Vauxhall Bridge, just half a mile from the Palace of Westminster. The company had long known how to tread softly through the corridors of power and on whose door to knock (or barge through). Now they could loom over everyone. Yet the Vickers Tower – known today as Millbank Tower – would enjoy its lofty record for just one year. The GPO Tower (now the BT Tower) was completed in 1964 and was 200 feet taller.

The strikingly modern design of the Vickers Tower did attract admirers. Staff and their children were delighted when Tony Curtis and Roger Moore came there to film an episode of *The Persuaders!*, the popular 1970s TV series and, in an ironic twist, a few months before the *Pisces III* incident the tower had been used as the set for a British film, *The Vault of Horror*, with Tom Baker and Terry Thomas among the characters trapped in one of the tower's lifts.

The chairman's office was on the south-east corner of the 29th floor, picked on the principle that in London the best room of a gentleman's residence should face south. Had convention been set aside and the north-east corner been chosen, the chairman could have gazed down towards the splendour of the Houses of Parliament, Lambeth Palace, over to the West End, then up to the green canvas of Hampstead Heath. Instead his office looked out onto the monumental chimneys of Battersea Power Station.

In the late summer of 1973, in whichever direction management looked there were storm clouds. As a nation Britain was in steep decline. The words of Dean Acheson, the US secretary of state, that 'Great Britain has lost an Empire but not yet found a role' continued to echo out loud and clear a decade after they were uttered. Industrial unrest was on the rise. The nation's miners had gone on strike the previous year for the first time since the 1920s, plunging the country into rolling power cuts, and despite a 20 per cent pay rise, their union leaders were eyeing the autumn and the opportunity of the cold winter months to repeat a painful process. The terrorism of the IRA was no longer confined to Northern Ireland. The events of 'Bloody Sunday' on 30 January 1972, when soldiers from the Parachute Regiment had shot dead 13 unarmed civilians in Derry, had worsened the conflict, as had the introduction of internment – the arrest without trial of suspected terrorists – with thousands picked up in dawn raids. Bombs had already been planted in London pubs, and more would follow in the autumn in Guildford and Birmingham.

Since the early spring the Vickers group had fallen within the sniper scope of David Rowland, a young financier from a family of scrap-metal merchants, with an unenviable reputation as an asset-stripper. Rowland had set up his own investment group at the age of 20, sold out for £2.4 million five years later and moved to Monte Carlo, from where he was now quietly buying up blocks of Vickers stocks. As the business pages of the *Guardian* reported, 'Within Vickers the attitude could best be described as one of uneasy calm.'

In the boardroom, which had a curved inner wall panelled in walnut featuring carved silhouettes of Vickers' most famous ships, the discussion was also centred around another more rapacious 'asset stripper'. The Labour Party was now in opposition, but their prospects of returning to power improved with every month of the Conservative premiership of Edward Heath, and Tony Benn, the shadow secretary of state for trade and industry, was vocal about the party's plans to nationalise aircraft manufacturing and ship-building, two of the pillars upon which the company's fortunes stood.

Vickers was described as 'the blacksmith and armourer to the nation', a company without whose might the fortunes of Britain in the Second World War might have been very different. What had started as a steel foundry, set up in Sheffield in 1828 by a miller called Edward Vickers, grew from casting church bells for village rectors to constructing the most complicated and destructive weapons of war yet seen. Over the course of the second half of the 19th century the company expanded from bells to marine driveshafts and ship propellers. Vickers moved first into shipbuilding, then armaments, producing its first sheet of armour plating in 1888, followed by artillery pieces two years later. The company's reputation as the nation's armourer began in 1897 with the purchase of the Barrow Shipbuilding Company and its subsidiary, the Maxim Nordenfelt Gun and Ammunitions Company. Sir Hiram Maxim, the American inventor of the machine gun, worked on a new design that was to become the Vickers machine gun, the standard issue for soldiers across the British Empire for the next 50 years.

The newly named 'Naval Construction Yard' in Barrow would become the company's spiritual home. In 1901 the firm produced the *Holland 1* submarine, the first of over 300 submarines to be launched there over the next 70 years. In

1911 Vickers expanded into aircraft manufacturing, and during the First World War both shipyards and aircraft assembly lines were yoked to the service of the nation. In the decade after the First World War, Vickers took over Sir W. G. Armstrong Whitworth & Co., an engineering conglomerate and rival shipyard, to form Vickers-Armstrongs, which could now boast a major shipbuilding yard on both coasts of northern England. By 1935 it was the third largest manufacturing employer in Britain, and the outbreak of war in 1939 would push every employee to the limit, for the company's ability to swiftly move into mass production would be the difference between victory and defeat during the months when Britain stood alone in Europe against Hitler's Nazi Germany.

During the course of the Second World War, Vickers-Armstrongs produced 50 per cent of the nation's machine guns and 75 per cent of its field artillery. The shipyards at Barrow and Tynecastle built and then launched 146 submarines, 22 destroyers, nine aircraft carriers, nine escort vessels, five cruisers, one monitor, one battleship and sundry merchant ships and transport vessels. The company's aircraft plants designed both the Spitfire and the Wellington bomber, then rolled out 20,000 of the former and 11,000 of the latter. Barnes Wallis, an engineer at Vickers, invented the 'bouncing bomb' that breached the Möhne and Edersee dams in 1943, while the engineering division produced thousands of tanks and guns for both the British Army and Royal Navy.

The post-war years were difficult, as a company on the scale of Vickers became a political pawn for both Labour and Conservative governments, pushed back and forth in support of their respective arguments over the merits and demerits of a state-run economy. After the 1945 Labour general election landslide, the English Steel Corporation owned by Vickers was nationalised, but when the Conservatives returned to power in 1951 the English Steel Corporation was privatised. Yet Vickers remained a company of unique national significance, sitting at the centre of the British defence industry. In the 1960s the Vickers yard at Barrow completed HMS *Dreadnought*, Britain's first nuclear-powered submarine.

Yet there was a need to diversify. Sir Leonard Redshaw, chairman and chief executive of Vickers Shipbuilding Limited, may not have conceived of Vickers Oceanics, but as the company's official biographer noted, 'He guided the pregnancy and made sure that a healthy child was born and grew up healthily.' Keen to seek new avenues for the skills developed in Barrow, Redshaw had noted the discovery of oil in the North Sea and recognised a new emerging market for his industry. In 1968, hearing of the success of a new submarine built by Hyco International Hydrodynamics of North Vancouver, Vickers made a financial arrangement with the Canadian company founded by Al Trice to buy one.

At the time, neither the Board of Trade of the British government nor Lloyd's Register of Shipping had any specific requirements for the operation of a miniature submarine. The technology was simply too new and government legislation had yet to catch up. So when the order was placed with Hyco for the construction of the craft it was agreed that it would be built in accordance with the guidance rules of the American Bureau of Shipping. The submersible, called *Pisces II*, was delivered in June 1969, one month before the moon landing. *Pisces II* was paired with a ship, Vickers *Venturer*, a 600-ton vessel, and together they offered a commercial submersible service capable of operating in sea state 4 (4- to 8-foot waves) and to a depth of 3,000 feet. Vickers' initial contract was with the Ministry of Defence for the retrieval of torpedoes from the waters around the military test range off the west coast of Scotland.

Yet this submersible service was never a major feature of the company. In 1970 the oceanics sub-division warranted a single sentence in the Vickers annual report:

Other activities included developments in the oceanics field, where the *Pisces* submersible has been employed on work for the Ministry of Defence, fisheries, geological institute and the National Environmental Research Council.

The following year the annual report was even more terse:

In oceanics, the *Pisces* submersible was in increasing demand for underwater search and research and a second submersible was accordingly put into commission.

After three years operating under the umbrella of Vickers Shipbuilding, Vickers Oceanics Limited was formed as a separate business in March 1972. Vickers held 63 per cent of the shares, the National Research Development Corporation 26.5 per cent and James Fishers & Sons, a shipping company that provided the support vessels, held the remaining 10.5 per cent. The new company was spearheaded by the introduction of Vickers *Voyager*, a larger ship of 2,850 tons, capable of carrying and servicing two submersibles: *Pisces III*, a sister boat to *Pisces II*, and *Pisces I*, whose diving depth was restricted to 1,100 feet. As a larger vessel, *Voyager* gave the company greater versatility as she was capable of operating submersibles in rougher sea conditions up to sea state 6 (13- to 20-foot waves).

The initial contract for Vickers Oceanics came from the Ministry of Defence, again to retrieve used torpedoes off the west coast of Scotland, but in 1973 the British Post Office hired Oceanics at the rate of £4,000 per day for a long summer contract to assist in the burial of the new transatlantic telecommunications cable. In the preceding ten years, between 1962 and 1972, the number of telephone calls between Britain and North America – both Canada and the United States – rose from 600,000 a year, or fewer than 2,000 a day, to 5.5 million – over 15,000 a day. In 1961 a new lightweight cable had been laid down from Oban in Scotland to Corner Brook in Newfoundland, known as CANTAT-1 (Canadian Transatlantic Cable); this was now to be supplemented, and eventually replaced, by a new cable – CANTAT-2. This new cable, manufactured by Standard Telephones and Cables, based in Southampton, was at the cutting edge of technology; Iain Finlayson, the submarine superintendent at the Post Office, described it as 'radio in a pipe'.

Designed to be one-fifth the size of previous transatlantic cables used by the likes of Bell Systems, while at the same time capable of carrying five times the number of calls, the new CANTAT-2 was only 1¾ inches in diameter and consisted of a central copper conductor surrounded by an aluminium tube, later insulated in polyethylene, which transmitted a wide band of radio frequencies stretching from 312 kHz to 13.7 MHz, broad enough to include the majority of AM broadcast and shortwave bands. The system worked by dividing the frequencies into 3,680 narrow channels, the low frequency carrying speech from Britain to Canada and the high frequencies carrying speech back to Britain, thus allowing 1,840 two-way telephone conversations to take place simultaneously. Every six miles along the entire length of the cable would be a repeater, 473 in total, designed to amplify the signal a couple of thousand times. The automated machine at Standard Telephones and Cables could produce 37 feet of finished cable per minute but was required to run 24 hours a day for almost a full year to produce the 2,800 miles required to span the Atlantic.

The Post Office hired the *Mercury*, at 8,962 tons the world's largest and most advanced cable ship, to lay CANTAT-2. Work started in Britain in June 1973, with the aim of completing the operation by December and the cable being fully operational for the connection of the first call by spring 1974. Meanwhile the *John Cabot*, a Canadian Coast Guard icebreaker, was hired for late winter 1973 to cover the last 170 miles into Halifax while hauling a 'plow' – a 17-ton, 24-foot-long structure with chisel blades that was designed to dig a trench four inches wide and two feet deep. The cable, and Vickers Oceanics involvement, were such big news that a reporter from the American magazine *Popular Science* was invited on a trial test dive to the bottom of the English Channel in *Pisces III*. The reporter was accompanied by Des D'Arcy, the company's chief pilot, who told him: 'Here's the emergency button. If anything happens to me when we're below, you just push it. Then open these two buoyancy valves and we'll surface in a hurry.' This was a conversation the reporter was later to reflect on, given what was to follow.

At least one member of Vickers Oceanics had endured the unpleasant experience of a serious submarine accident. In 1972 Vickers Oceanics appointed a new managing director, Commander Peter Messervy, whose task was to increase both profitability and safety by bringing in a greater degree of discipline to a company operating in a frontier industry and staffed by a mixture of mavericks and oddballs.

Messervy had a cool head, a steady hand and a brave heart, as well as a George Medal to prove all three. Standing just over five feet eight, with broad shoulders, a thick neck and the bashed nose of a boxer, he had a commanding presence garnered from 35 years' service in the Royal Navy. Cut him – at your own considerable risk – and he would bleed salt water. In 1935, at the age of 15, he had signed as a boy seaman based first at HMS *Ganges*, the navy's training facility in Harwich, before going on to serve on the battleships HMS *Resolution* and *Royal Sovereign*. He spent three years of the Second World War barking orders at young recruits as a physical training instructor before landing at HMS *Elfin*, the submarine base at Blyth.

The postwar years saw him slowly clamber up the ranks, like an unfit recruit on a knotted rope: warrant officer by 1945, then commissioned bosun by 1949, the same year he made a name for himself in the boxing ring. An avid and enthusiastic pugilist, Messervy fought for the light heavyweight championship at Wembley Arena in 1949, but in a curious twist both he and his opponent were dismissed from the fight for 'not giving their best'. Angered at the decision, Messervy fought back, secured a return bout the next year and succeeded in taking the belt.

During the Korean War in the 1950s, Messervy served as a senior commissioned bosun on the aircraft carrier HMS *Unicorn*, before reaching the conclusion that he would have to go underwater if he wanted to advance further. He retrained as a diver at HMS *Vernon* in Portsmouth, which specialised in torpedoes and mine-laying, and as a frogman he would find his natural forte. He liked the stillness of the underwater world and the absolute focus required for repairs, recoveries or, his favourite task, defusing sunken munitions. He also liked the camaraderie that came with operating in a smaller team with sharper skills, skills he was only too keen to demonstrate on his first posting in Scotland. Based at the Royal Navy's base at South Queensferry on the Firth of Forth, Messervy and his team were charged with removing live mines from the wreck of the *Port Napier*, a minelayer that had sunk in Loch Alsh in 1940 when a fire broke out on the ship.

In the small pond of experienced divers, Messervy was a big fish, one that grew in size and weight on being appointed to his first command in the summer of 1959, when he took over as Far East Fleet Bomb and Mine Disposal Officer, based at HMS *Terror*, Singapore's magnificently titled naval base. The waters around the island of Singapore and its harbour were teeming with sunken munitions, a legacy of the unsuccessful defence against the Japanese during the Second World War. But there were also natural dangers. Five years earlier, in July 1954, a young naval diver called Charles Brian Larkin, aged 21, was diving in 20 feet of water close to the city's commercial harbour when he was attacked by a reef shark. Although rescued by his colleagues and dragged back onto the dive boat, he died from his wounds. When Messervy was told the story, his attitude was, 'Fuck the sharks.' He believed that dangers could be mitigated but never avoided, and at times should be resolutely endured.

In Singapore Messervy faced his biggest challenge to date. He and his team were tasked with removing six live torpedoes from the hull of *I.30*, a 3,000-ton Japanese submarine that had been sailing off the coast of Keppel Island in October 1942 when she triggered a mine and sank in the explosion. Japanese divers had recovered a stash of German weapons and various pieces of instrumentation but had abandoned plans to recover the torpedoes as too dangerous. For weeks Messervy and his divers descended to a depth of 42 feet and worked for hours in, at times, zero visibility, cautiously wielding an acetylene torch to cut through steel, only two feet from the live torpedoes.

In the middle of the operation Messervy received a call from the British Army. On a training exercise in the Johor jungle, a soldier had clumsily dropped a secret and newly developed rifle designed by the Belgian government, and it now lay at the bottom of a river. Anxious that it didn't fall into the wrong hands and its novel design be revealed, the army called in Messervy, who succeeded in recovering the rifle in a single dive before returning immediately to the site of *I.30*. For the successful completion of the torpedoes recovery, Messervy was awarded the highest medal for bravery in peacetime, the George Medal.

Messervy rose to the rank of lieutenant commander and, after service with the anti-submarine warfare school, took command of the salvage vessel HMS *Reclaim*. It was while on *Reclaim* in the middle of the Irish Sea that Messervy first encountered a *Pisces* miniature submarine, which was being used to find an Aer Lingus Viscount jet that had crashed, killing all 61 passengers and crew, in March 1968. Over 26 days, Messervy's team set an endurance record, making 91 dives to 250 feet to recover wreckage, but despite a two-year investigation, the cause of the crash was never comprehensively established.

Messervy retired from the navy in 1970 after 35 years' service, but was determined to carry on working at sea. Teaming up with two fellow naval veterans, they set up HMB Subwork, based in Harwich and designed to cash in on the gold rush heralded by the discovery of oil in the North Sea and the projected need for miniature submarines. It was while developing this business that Messervy was introduced to the T-shirt boys, but unfortunately also found himself stranded on the bottom of the sea at a depth of 600 feet in an immobilised miniature submarine: *Pisces III*.

In his research into miniature submarines, Messervy had travelled to Vancouver in 1971 to inspect Hyco's operation and train in piloting their latest submersible craft. Al Trice, who co-founded Hyco, and was now developing and marketing the submarines he'd helped to design, arranged to take Messervy out on the company's barge for a training dive at Indian Arm, a stretch of water close to Vancouver that had a maximum depth of 600 feet. The Jervis Inlet further up the coast dropped down to over 2,000 feet, so on this occasion

Indian Arm was the preferable and most fortunate choice, considering what was to follow.

Messervy was joined in the dive by Fred Warwick, an experienced pilot, and together they were lowered overboard and began a controlled descent, but Warwick immediately noticed that *Pisces III* was tail-heavy. There had been an oversight during the dive checks and no one had noticed that the aft sphere's vent plug was not in place; as a result, as soon as they dropped below the surface the aft sphere began to fill with water, rendering the submarine too heavy to raise itself from the bottom. Fortunately the Hyco barge had a second submarine on board, the *SDL*, which was undergoing its first sea trials. The *SDL* was sent down, piloted by Mike Macdonald and Lt Barry Ridgewell, with a heavy polypropylene rope attached to a lift hook wedged into the sub's mechanised talon. In a textbook rescue, Macdonald and Ridgewell were able to attach lift lines to *Pisces III*'s main lift point, and after four hours the sub was winched safely to the surface.

An error as potentially fatal as a missed vent plug would have led to a court martial under Messervy's command, but instead he was greeted by the benign smile of Al Trice, who insisted they had simply tried to make the training as realistic as possible. Messervy was not amused – as he later told his wife, to be trapped in an immobile submarine at great depth, unable to assist in one's own rescue, was 'a terrifying experience'.

CHAPTER FOUR

When Roger Mallinson was ten years old, his headteacher tried once again to beat some sense into him, but for Roger it was one time too often. As a pupil at the local primary school, a simple whitewashed building in the Cumbrian village of Winster on the southern edge of the Lake District, Mallinson was used to being sent to see the school disciplinarian and custodian of 'Tommy Tickle', as the cane was called.

As it happened, the headteacher was also his mother – Doris – and any hope of leniency evaporated when he was first caned at her hands, to be replaced by fear for all subsequent appointments. 'She really used it with venom,' he recalls. Yet on this occasion young Roger rebelled. Doris reached for the cane, but before she could swing it back, her son snatched it from her and broke it over his knee. She then opened her desk drawer and brought out a 12-inch wooden ruler. 'I had broken "Tommy Tickle",' Mallinson says, 'so I could break a ruler. Every ruler she tried to hit me with, I broke.'

For Doris, a moment of clarity came when she found herself struggling to open a new packet of rulers. She laid the packet down, then dismissed her son, but not back to the classroom this time but to an anteroom where he spent the remainder of the day. She'd had enough. He never returned to her school, but instead continued his education at a neighbouring school, St Mary's. At the age of ten, Mallinson learned two things: that the only way to deal with a bully, even one's own mother, was to stand up to them, and that his future lay not with books but with crafts. He could no longer bear being beaten across his hands, or as he called them, 'my working tools'.

Doris Mallinson had given birth to twin boys, Roger and Miles, when she was forty, and, according to Roger, never forgave the pair of them for their surprise appearance in her life. A dedicated and ambitious teacher, Doris was disappointed that Roger possessed little aptitude for books, barely making it past the first few lines of the opening page of any book he read. But she was blind to his practical ingenuity. Despite his difficulties with his mother, Roger would describe his childhood during the Second World War as 'advantaged'. In Miles he had a best friend and partner in ingenuity, and together they would work on the attic floor until the early hours of the morning, 'four hands and four eyes and two brains all working on the same thing'.

Then there was wartime Britain, an age of austerity but also opportunity – 'We had a fabulous time,' Roger recalls. 'We couldn't have been born at a better time. There was nothing in the shops. There was no money. If you wanted something you had two choices: you either did without or you made it yourself.'

The boys' father, and Doris's husband, was Edward Mallinson, a skilled engineer and a native of the Lakes, born in the Oakland area of Windermere. During the First World War the family lost both an uncle on the first day of the battle of the Somme and, later, the family home. In June 1916 Edward, then a 14-year-old, had been sent to the butcher's for a joint of meat, only to have a well-informed neighbour, possibly one who'd been in previous conversation with the telegram officer, shout, 'Hey, Teddy, will you tell your dad your uncle Jack was killed.' The death of Edward's father in 1918 meant the loss of the family home – which had been tied to his employment – and the eviction of the family. A hatred of the class system and the motives of the moneyed were handed down by Edward Mallinson to his son, but he also instilled in him a love of industry and mechanical design.

During the Second World War Edward Mallinson was absent from home, toiling in the manufacturing yards of Vickers Shipbuilding, where he worked on submarines. An accident in childhood had left him partially deaf, and when he failed to hear an air-raid siren at Vickers he was injured by falling rubble after the works took a direct hit from German bombs. Roger, although obviously concerned for his father, was delighted that he returned to Windermere to recuperate, and upon his recovery he went to work in a nearby factory manufacturing Short Sunderlands, the flying-boat patrol bomber. A cherished childhood memory of Roger's was attending the Christmas party and being allowed inside the workshops to witness the flying boats being born. It was the smell he loved the most: a pungent perfume of metal, engine oil and grease. Afterwards he began to design and build his own model flying boats.

At Windermere Grammar School, Roger enjoyed wood-work, chemistry and physics, but struggled with mathematics. The headteacher, who also doubled as the maths teacher, insisted on wearing a mortar board, gown and hood, and routinely arrived at class ten minutes late to demonstrate his authority. Yet he was dismissive of his pupils. During a lesson on trigonometry he declared: 'Some of you will learn, some of you will not and those who will not, never will.' Mallinson believed he would be among those who 'never will'.

At the age of 16 Mallinson and his brother left school, and their father secured apprenticeships for them at Vickers-Armstrongs in Barrow. These spanned five years, during which Mallinson would pass through a range of different departments, beginning with the drawing office to learn the skills of technical drawing, before moving into process planning and later manufacturing. A couple of years into the apprenticeship, he was reintroduced to the lessons of the school room but in a manner that made sense. While working on a large combination turret lathe, the technical drawing specified an angle of so many degrees and so many minutes and so many seconds. Mallinson had never seen anything like this. He asked the foreman, who measured just over four foot tall and was dressed in a little faun smock coat, and he replied: 'Oh, it's easy enough – it's trigonometry.' Over the next hour, he initiated Mallinson into the mnemonic of 'Peter's Horse Brings Home Peter's Bread', to help him learn about sine, cosine and tangent. He then walked him over to the sine tables that ran along the walls of the workshop in five-inch letters and which had so far been meaningless to Mallinson.

He explained, 'Suddenly, if you looked at the angle you wanted, you found the number of the sine, then multiply that by ten and go to the stores and get a sine bar and some block gauges for sine angle times ten. Bring them back to the machine, but put the sine bar against the angle you want to put on, then put the block gauges at one end of the ten-inch sine bar then put the other end of the sine bar onto the table and then pull everything together and then you will get the angle.' Simple.

At the end of his five-year apprenticeship Mallinson decided it was time to move on. To celebrate the completion of his apprenticeship Mallinson bought his first – and to date only – car: a mint-green Austin 7. On the date of purchase, 11 October 1958, the vehicle was already 27 years old. He paid £25 for what he described as 'a little cracker', which can't be far from the truth: by 2020 it had clocked up over a million miles. The first lengthy journey they took together was to London, as Mallinson had got a job nearby in Croydon with a company that specialised in punch-card accounting machines. By then he was already dating his future wife, Pamela, but the couple had no plans to move south, so Mallinson commuted every week for over a decade. He would set off on Sunday night and on arrival top up his car at the same garage on Edgware Road, lest it run out of petrol on Park Lane. Although Mallinson liked his colleagues, he had, like many northerners, a suspicion of London, where he thought that 'everyone fights over their own importance.'

Mallinson may have spent the swinging 60s in London, but it had little effect on his craggy northern intransigence. When the company he worked for, which had amalgamated a number of times during his tenure and was now known as Powers-Samas, offered him a choice of transfer – Accrington or Aberdeen, neither of which appealed – Mallinson preferred to resign and head home. Now the father of three children, Mallinson was also in the process of designing and building a family home in his native Windermere: 'I was a Windermere man and I thought: "Sod this!"'

After a few weeks back home, he decided to visit his old firm Vickers to ask what jobs were going. He was told the company had recently launched a sister venture, the aforementioned Vickers Oceanics, and he was duly dispatched for an interview in which although both 'boats' and 'ships' were mentioned, he was never once told about the type of vessel on which he would soon be trained. It was only when he'd been offered the position and given a guided tour around the facility that Mallinson was shown the brick building referred to as the 'loco shed' – where railway engines were once repaired – and saw the vehicle for which he had been hired as 'pilot'. Peering through the open doors, he glimpsed the red sheen of what, he later learned, was *Pisces II*. 'That was the first time I'd seen it,' he recalls, 'and the first time anyone had mentioned the word "submarine".'

Mallinson began in October 1971 and was violently sick during the first few days. Later, when asked why he had applied for a job piloting a submarine despite never having been to sea, he replied: 'You've got to start somewhere.' So where Mallinson started was as a submarine pilot off the coast of Oban on the west coast of Scotland. His instructor was an enthusiastic smoker who did not believe that spending eight to ten hours in a sealed space was cause for kicking the habit or even cutting down a little. Instead he proceeded to chain-smoke for the entirety of the dive, explaining to Mallinson that this was on the grounds that his lighter didn't work at depth, and once a cigarette was extinguished, that, unfortunately, was that.

Mallinson would become enchanted by his new work environment:

> It was absolutely fabulous. I had never been underwater and there you were with a thousand watts of light, and you could see all the fish and plant life on the bottom. It was like looking into an aquarium. The fish played with you, especially at night, when there was no other light. It was almost impossible to do your job with so many fish around.

Vickers Oceanics' principal contract was with the Ministry of Defence, retrieving torpedoes and mines for the Royal Navy. This was frequently from the torpedo range off the Kyle of Lochalsh and between the small islands of Raasay and Rhona, where the water was well over 1,000 feet deep. The torpedoes were equipped with 'pingers', and were found either buried in the soft silty sand or standing up on their propellers. The task of the miniature sub and crew was to find the torpedo's centre of gravity, clamp on the mechanical grip and then bring it to the surface. The most problematic job was when the torpedo struck an underwater wall of rock and broke in two.

Over time Mallinson became expert in tracking and retrieving torpedoes. He took pride in trumping the officers of the Royal Navy, whose manner at times he found condescending. When a torpedo was lost in the Sound of Raasay, the navy spent five months attempting to locate it themselves before calling in Vickers Oceanics, whose contract had by that time lapsed. This delay caused the pinger to begin to fade in volume, yet Mallinson was determined to succeed all on his own, where the might of the navy had failed. In the grey choppy waters of the Sound, he announced that he wouldn't surface without it, then closed the hatch with a theatrical flourish. When he touched down, he contacted the surface to say, 'On bottom', but was determined thereafter to maintain radio silence until he had brought the matter to a successful conclusion.

In the dark, brackish depths of the Sound, Mallinson trained the lights on the seabed, searching for the remnants of any skid marks left by the torpedo. He then took off his headphones, turned off the underwater telephone, switched off the gyrocompass, silenced anything that made the slightest noise and simply listened. The pinger was on a very high frequency that you could hear if you listened carefully for its bat-like squeak. Mallinson picked up the first faint sound, then inched the sub forward, ever conscious of the pinger's rise and fall. In time he discovered the torpedo buried in mud, and after clearing it up and securing it, he released only the second message of the dive: 'Torpedo in claw. Ready to ascend.'

Mallinson was to take a swift dislike to Peter Messervy, the man who would lead his rescue. First there was Messervy's manner, which was brisk to the point of rudeness, and then there was the style in which Messervy wished to be addressed. All staff were expected to refer to him as 'Commander'. Mallinson felt that as he himself was not in the Royal Navy and, frankly, no longer was Messervy, such ranks were inappropriate, yet the management at Vickers Oceanics were only too happy to pander to Messervy's whims. Other staff, too, were expected to fall in line. The fact that Messervy wore a monocle was also a matter of intense irritation to Mallinson, who viewed it as the height of affectation and the mark of a braying toff. Messervy's management style didn't help matters either. On one occasion Mallinson spent three days and nights slaving around the clock on a refit of *Pisces III* after work had fallen behind schedule. Once the task was completed he took a longer lunch break than usual in order to visit his brother Miles, only to be reprimanded on his return by Messervy for being 'late'.

Then there was the time Messervy asked to see Mallinson and suggested that the company should make 25 gold models of the sub. Intrigued by the challenge, Mallinson, who had never worked with gold, persuaded his local dentist to help him build the moulds using dental clay. But when he approached Messervy to pay for the work carried out so far, the Commander changed his mind about the project and refused. Unwilling to see the design go to waste, Mallinson covered the costs himself and brought in a gold ornament from home, which was melted down and poured into one of the moulds. As Mallinson recalls, 'He was a pillock.'

* * *

When Roger Chapman was 15 years old he was introduced to his first submarine. The son of a Royal Navy commander, a life spent on the ocean waves seemed predestined, but no one had yet suggested a life and career spent under the waves. On a visit to Gosport in Hampshire for a two-day interview with the navy, Chapman and the other six naval-minded schoolboys in his party were taken on a tour of HMS *Dolphin*, headquarters of the British Submarine Command. It was an evening visit and, en route, a precocious pupil had pestered the chief petty officer in charge to stop for a pint, so it was in a state of mild intoxication that Chapman fell in love.

HMS *Totem* was an out-of-service T-class conventional submarine. As Chapman later wrote:

> All I can remember is that the compartments looked incredibly small and cramped, while there was a strong smell of fuel oil. At the time we put this down to the fact that this particular submarine had been out of commission for some time; but as submariners will know, all diesel-driven submarines have this distinct smell, which clings to clothes and personal belongings long after you leave the confines of the hull.

Young Roger was marked as if by perfume.

To the expectation of his father and resignation of his mother, on his 18th birthday Chapman joined the Royal Navy at Dartmouth. In 1963, during the second year of his training, he sailed with a naval frigate on a training exercise in the South China Sea. The frigate was tasked with hunting down the submarine HMS *Anchorite*. After the first part of the exercise was completed, while both vessels were berthed in Hong Kong harbour, Chapman arranged an invitation to board *Anchorite* and succeeded in gaining access to the ward-room. With his preternatural confidence, he decided to ask if he could hitch a ride with the sub when the fleet sailed from Hong Kong to Singapore for the next leg of the exercise. As an officer had already flown on to Singapore, there was a spare berth and, as a potential recruit to the 'Silent Service', Chapman was invited aboard.

The voyage to Singapore took five days, and although *Anchorite* never once submerged, sailing instead on the surface, Chapman felt that he could have been at a depth of 3,000 feet since all the hatches on the submarine were closed. For five days he was immersed in the life of a submariner, and his future course was assured. Shadowing senior officers as they went about their tasks, he gained an insight into how submariners were set apart from the rest of the navy. 'In a disciplined service such as the Royal Navy,' he later said, 'one of the biggest attractions of submarines is the apparently relaxed atmosphere once at sea, combined with the general respect officers and men have for each other while living and working together in confined spaces.'

He recognised a heightened sense of responsibility among each member of the crew, forged from the knowledge that so precarious was their environment – thousands of feet below the surface and often thousands of miles from home – that even simple errors could have catastrophic consequences. These psychological pressures created a unique band of brothers, to which he now wished to belong. Chapman was accepted into the submarine school at HMS *Dolphin* and would spend the next four years in conventional submarines, 'mainly East of Suez', as he explained in the imperial jargon of the time.

In the summer of 1969 love first surfaced in his life. June Sansom was a Yorkshire girl, with ambitions beyond its borders. Educated at a secretarial school in Edinburgh, she subsequently moved to London with a group of friends, and it was at a house party in Islington, to the soundtrack of 'Cracklin' Rosie' by Neil Diamond, that she first met Chapman, who was on leave at the time. He was 24 to her 20, and they soon fell in love. A perk of the naval profession was access to yachts kept in the marina near naval headquarters at Gosport, and during the summer Chapman and his fellow officers would take their dates on sailing outings through the Solent and around the Isle of Wight.

June was not a natural sailor, and it was an escapade highlighting her inexperience and vulnerability at sea that helped nudge Chapman towards a proposal of marriage. While Chapman was on duty, June had gone on what was planned to be a short sailing holiday to Jersey with her brother and some friends. As the weather had turned foul, they decided to leave the boat in Guernsey and instead fly to Jersey. Later, after collecting the yacht and starting to sail home, the party was caught in a second streak of bad weather. As the yacht rolled in heavy waves, June became violently sick. So dangerous were the conditions that they were forced to abandon the boat and accept a rescue from a passing oil tanker. Although June had packed the yacht with her worldly possessions as she was moving flat, she was not allowed to take anything on board the tanker, not even her handbag. The yacht was abandoned mid-Channel, and June and her party were kindly dropped off at Southampton. (The boat was later found on a beach with her possessions intact.) The idea of June in peril at sea prompted Chapman to reassess the true depth of his feelings, and this led to his proposal of marriage later that summer.

In August 1970 the newly engaged couple prepared to go their separate ways for a while, Chapman on a round-the-world trip, exact destinations and dates classified, while June, not one to wait wistfully at home, had accepted a job in Paris as a secretary in the foreign exchange of the French investment bank, Generale Occidental, run by the buccaneering capitalist James 'Jimmy' Goldsmith. 'He was a lovely man, a complete gentleman,' recalls June of Goldsmith, although in 1974 Goldsmith and fellow members of the Clermont Set, a private group of gamblers based at the Clermont Club in Mayfair, would be suspected of assisting in the disappearance of Lord Lucan, the prime suspect in the murder of his nanny, Sandra Rivett.

As it happened it was June, not her fiancé, who did the most travelling that first year of their engagement. Engineering faults in his submarine meant Chapman spent most of the year stationed in a naval dockyard. Concerned lest his betrothed be wooed by the Gallic charms of a Parisian, he wrote her regular letters but in this he was stymied, as was the rest of the United Kingdom, by Post Office workers, who in January 1971 began a seven-week strike in demand of a pay rise. Chapman, determined not to be foiled by industrial action, attended an international rugby match between

England and France at Twickenham and pressed his letters on a Parisian fan, who duly posted them on his return home. Although the gesture delighted June, it did leave her to at first suspect that he had secretly moved to Paris, as the postmark carried a stamp of the Louvre.

June returned from Paris in August 1971 and the couple were married in October at her home village of Bolton-on-Swale. Chapman had been posted to Barrow, where he was to join the crew of HMS *Swiftsure*, a new nuclear-powered submarine, then under construction at Vickers. As there was no official naval accommodation in the area, the couple rented a small cottage in Great Urswick, near Ulverston. For June, that autumn was awful as her husband was away and the weather was wet. When Penny Old, a fellow naval 'widow', called to invite her round for tea, June dropped everything and was on Old's doorstep half an hour later. By January, when Chapman was due to depart on a three-month commission, June was so miserable that she returned to London, to the company of her old flatmates and a round of temp work.

For Chapman, his next commission was to have lasting repercussions. As *Swiftsure*, the lead ship of her class of hunter-killer subs, was not yet complete, he had taken the opportunity to serve a temporary role in a sister sub, completed two years previously and now undergoing extensive sea trials. The navigator had been shipped home with a double hernia and Chapman would serve as replacement. During his deployment Chapman began to have doubts about the quality of his eyesight. Submarine service requires almost perfect vision, yet at the time of his admission the standards had been lowered slightly to ensure a sufficient intake. Yet during a routine medical and a subsequent eye test, it was discovered that his sight had further deteriorated and was now below the permissible level for submarine service.

The question now was whether he should continue to serve but exclusively on surface vessels, one that he was reluctant to answer. Service on submarines often spoils a mariner for any other type of vessel, and although he knew his father would be keen for him to continue in the Royal Navy, he was no longer so sure he wanted to. Did he really wish to return to surface fleets, and undergo a nervous annual eye test to determine whether further deterioration would mean enforced retirement? 'Alternatively,' he later recalled, 'I could leave straightaway and embark on a new career in the outside world.' Chapman chose the world.

For the next few months Chapman returned to Barrow and HMS *Swiftsure*, the submarine on which he would still serve and sail, waiting for his exit papers to be processed. The sea trials for *Swiftsure* were exhilarating but to Chapman bittersweet. Built at a cost of £37.1 million, *Swiftsure* was the tip of a new spear of nuclear-powered submarines; although not armed with nuclear weapons, she was equipped with five torpedo tubes and an arsenal of Spearfish torpedoes and Harpoon missiles. She would become quietly famous in the Silent Service for successfully tracking down a Soviet military exercise, slipping undetected beneath the aircraft carrier *Kiev* and 'stealing' its acoustic signature. According to reports never confirmed, *Swiftsure*, while submerged, was able to raise its periscope to within ten feet of *Kiev*'s hull and take photographs of its propulsion system. For Chapman these would be distant adventures on which he would not sail, although he did participate on trials in which *Swiftsure* ran deeper and faster than any previous Royal Navy submarine.

Prior to his departure in the autumn of 1972, Chapman attended a number of job interviews in London, but he eventually found a position closer to home. The offices of Vickers Oceanics were only 300 yards from where *Swiftsure* lay in dock, and when Peter Messervy heard that Chapman was leaving the service, he invited him for an interview in the Portakabin kingdom he had rapidly constructed, an upgrade from the small static caravan that was his first 'office'.

The vacant position was for a pilot, but Messervy made it clear that he expected more from Chapman. Vickers Oceanics was, Messervy believed, a cowboy operation in which staff rode roughshod over what few rules and regulations currently existed. These insubordinate employees needed to be smartened up or weeded out, and what was required was a spot of naval discipline instilled by a fresh cadre of military men, hard workers with cool heads who would drive the company to greater success while ensuring that safety standards were maintained. Chapman explained that his eyesight had prompted his early retirement but Messervy insisted his vision would be more than adequate for commercial work. For Chapman, the position seemed perfect, as he and June could stay in Barrow and he could return to the undersea world he so loved.

There was even a job for June. In the past she had found it impossible to obtain a local position in the area as employers were reluctant to hire a naval wife who was bound to be moved at short notice to her husband's next posting, but this was now no longer the case. She was hired by Vickers Oceanics' parent company, and in November 1972 both started on the same day, with Roger driving them to work in the couple's second-hand Volkswagen.

June's job was as a secretary in Vickers' Development Projects Team (VDPT), based on the third floor of a brown breeze-block building down by the dockyard. The office was open plan and she was one of three female secretaries working for ten male managers. The women were fenced off in a small corral by rows of filing cabinets that acted as a partition, and they'd sit there from 8.30 till 5.10 – with a 65-minute break for lunch – hacking away at their heavy manual typewriters. All correspondence required five copies with carbons, so you really had to hammer on the keys. This noise became the soundtrack to the day – that, and laughter, for the third floor was a fun place to work. Never before had she worked in a company with such strong unions, certainly not under Jimmy Goldsmith at Generale Occidental, and every few months seemed to bring an extra bulge in her pay packet, paid weekly in cash in a neat brown envelope.

She worked directly for the head of the department, George Henson, clean shaven with dark swept-back hair, always smartly dressed in a black suit, white shirt and navy blue tie. Billows of pipe smoke would drift up from his desk and June regularly had to pop out for his favourite tobacco. Henson was also a diabetic, and part of June's daily routine was reminding him mid-afternoon to drink a strong cup of tea with three or four sugars, lest his pallor turn grey. She got on well with the other girls – Joyce, who was dating one of the younger men on the staff, another whose husband was on dialysis and had to go home promptly – and no one minded covering for her if they had to work late on Fridays, when Henson preferred to get most of his work done.

In fact, it was one of the girls in the office who inadvertently told June that she would soon be moving into a new home. When the couple first visited Broom Cottage in Broughton-in-Furness, it was just before Christmas and raining heavily, and the drabness of the house, redolent of the 1950s and lacking even a basic kitchen, had put the couple off and they'd turned it down.

June, however, had doubts about their decision, thinking that with enough work it could have been a happy little home, so she was intrigued to know who had bought it at the lunchtime auction. One of her colleagues, with an eye on another property, had attended, and returned to describe a 'swarthy guy, with receding hairline and a navy blue sports coat'. When more details were given, the more familiar the buyer became, until June laughed and declared, 'You've just described my husband.' At least the price bid was under their budget. In the evening, as June climbed into the car and Chapman said, 'I've something to tell you,' she replied, with a mixture of delight and annoyance, 'I know.'

Broom Cottage was to be June's 'little project'. They didn't get the keys to the property or exchange contracts until July 1973, the height of the summer season, when the 'weather window' in both the North Sea and the Atlantic was wide open and Vickers Oceanics were at their busiest. The couple agreed that while Chapman was at sea on the new CANTAT-2 project, June would endeavour to turn the drab little cottage into a bright new home in which to raise the family they were anxious to start. All that was required was a little elbow grease, a fair amount of chipboard and gallons of magnolia paint. For both Roger and June there was an air of excitement about their new life.

PART II

WEDNESDAY

CHAPTER FIVE

There's a blackness so thick you can almost reach out and touch it, and from the blackness comes the keen whine of leaking air.

Mallinson and Chapman are disorientated. The ache of limbs clattered against sharp-angled equipment, even through the padding of a cushion, is a reminder that they are both alive, if in the gravest of danger. Mallinson thinks about the last reading on the depth gauge before the lights went out and knows they are so deep as to be surely beyond rescue, but any further dark thoughts are knocked out by the noise of the leak and the need to shut it off.

He recognises the sound of air escaping from one of the oxygen cylinders, but where is it? The exact configuration of *Pisces III* is not yet clear to her pilots. They remain in the dark, both literally and figuratively, and illumination isn't going to come quickly, not with the lights out and the batteries now in who knows what kind of condition.

But the batteries can wait. The renegade oxygen supply cannot.

Mallinson begins to feel around for the oxygen tank, recognising different pieces of equipment by their contours, and begins plotting their jumbled new location. It takes between two and three minutes before his hand lands on the cold steel case of the oxygen cylinder, finds the round control valve and shuts it down. The whine is reduced to a whisper, then silence. Both men can now hear themselves worry.

Chapman, meanwhile, has got his hands on the torch. He switches it on, summoning a cool white cone of light that acts as a guide to the disarray. Their small spherical world has been turned upside down. *Pisces III* seems to be standing on her tail, with the stern wedged deep into the Atlantic's sandy bottom, leaving the accommodation sphere standing up at right angles to its usual position. The two 'beds' on which they had laid while working on burying the cable are now like a pair of stretchers propped against a wall. The 'floor' they're standing on was, prior to the crash, the rear wall on which the sonar equipment, oxygen regulator, valves and underwater transducer were fixed. Each torch sweep reveals another scene of chaos from the collision, as well as glimpses of Mallinson's bearded, tense and worried face.

Mallinson is thinking about the oxygen bottle and how much has flowed into the atmosphere around them. While he knows that it's not 'lost', that they will breathe it in over the next few hours, he's also aware that it must be carefully rationed, eked out into this constricted new world in which they will have to live. How much longer will it last? This leads him to the battery and the power supply, as whatever oxygen supplies they do have will be rendered useless if they can't operate the scrubber and continually clear the carbon dioxide out of the atmosphere.

To power it they have to switch the batteries back on, and Mallinson is more worried about this task than he feared their initial plummet. His greatest dread is a fractured cable igniting an electrical fire. A fire inside the oxygen-rich sealed steel sphere of *Pisces III* would be unstoppable, the underwater equivalent of the flash fire inside the command module of Apollo I that burned three NASA astronauts to death six years ago. Those men had a ground crew struggling to save them. At almost 1,600 feet down, the only witnesses here will be fish, attracted to the light of a raging fire behind the porthole glass, a fatal brief flicker in the darkness of the deep.

Mallinson discusses the dangers with Chapman, but both know there's no choice. They have to turn on the batteries in ascending levels of voltage, starting with the 12-volt, then the 24-volt and finally the 120-volt, potentially the most lethal. Chapman shines the torch on the central control panel and Mallinson reaches out to flick the first switch, steeling himself as he does so.

The first switch operates smoothly.

Then the second.

If a deadly fire burns in their immediate future, it will come from switch three, when 120 volts surges through the cable. Mallinson flicks the switch. There's an audible click, and suddenly the sphere is bathed in light.

The batteries have survived the crash. The two of them have electrical power, at least for now, and a limited supply of air. The next crucial item on their checklist is the efficacy of the carbon dioxide scrubber. Exhaled with every breath, CO_2 is lethal if allowed to build up in a confined space. The operating procedure for *Pisces III* is that with two men breathing inside a steel sphere only 80 inches in diameter, the CO_2 level will quickly rise and so must be monitored by careful consultation of the Ringrose CO_2 indicator, a small thermometer-style device held inside a steel panel the size of a paperback novel, which is bolted on to the sphere's wall. The 'thermometer' indicates the level of CO_2 in the atmosphere, and regulations state that when it passes 0.5 per cent, which usually happens every 30 minutes or so, then the scrubber is activated.

The scrubber looks like a tapered steel plant pot with a tin of paint on top. The 'paint tin' is a canister of lithium hydroxide. The tapered steel plant pot into which it's inserted holds an electric fan, and once activated the fan draws the sphere's stale air into the lithium hydroxide, which binds with the CO_2 molecules, trapping them and thereby purifying the air. This takes between 10 and 15 minutes, and is monitored via the Ringrose indicator. The removed CO_2 must then be replaced with oxygen from one of two cylinders that, once activated, slowly 'bleeds' oxygen back into the atmosphere. The pilots know when to switch off the oxygen – when the atmospheric pressure returns to the initial reading taken and fixed when the sub's hatch is first closed at the beginning of the dive.

If switching on the batteries held the grim possibility of swift and violent immolation, switching on the scrubber provokes its own anxiety. If the scrubber doesn't work, they will still die this day.

The angle at which *Pisces III* has come to rest means the scrubber and its canister are lying horizontal, fixed by a bracket to the wall. Chapman reaches for the scrubber, turns it upright and looks over to Mallinson. Both men know exactly what's at stake.

'Try it,' says Mallinson.

Chapman presses the switch. The click is comforting, but not as much as the gentle purr as the motor and fan are roused back to life. A wave of relief rolls over both men, followed by a second wave of exhaustion that barrels into Chapman. Later he'd write:

> We had been up most of the previous day and night, and just a short while ago had been so near to a bath, breakfast and welcome rest. Then this. It was still almost impossible to understand our predicament, but my body gave the game away. I was shaking like a leaf and very hot.

The predicament is that they're trapped on the seabed and any attempt at self-rescue will be suicidal. If they were at a depth of 100 feet, they could hold their breath, allow the submarine to fill up with water until the sphere was entirely submerged and the pressure inside matched the pressure outside. This would then enable them to open the hatch. Even if this were possible now, at a depth of almost 1,600 feet, the weight of water pushing down on them would be the equivalent of 50 tons, and even if their organs didn't collapse and they could begin to swim up and away through the darkness, how long would it take to glide through water up the side of the Empire State Building and beyond? Ten minutes? Fifteen minutes? Twenty?

* * *

Oxygen consumption is a matter of imprecise arithmetic. A *Pisces* crew will, on average, consume one litre of oxygen per minute per man, recorded as '1 litre/min/man' and consumed via two oxygen cylinders, each carrying 63 cubic feet of oxygen and capable of taking a pressure of 3,000 psi. Under normal circumstances, two oxygen cylinders can expect to last 30 hours at the standard rate of consumption of 1 litre/min/man. In an emergency scenario, the crew are expected to lie down, reduce all physical action to the absolute minimum and restrict conversation to the bare essentials, all the time breathing in a slow and steady manner. The inhalation rate is then reduced to a target of 0.5 litre/min/man, the goal being to take a little over one day's oxygen supply and stretch it to two, maybe three days. In an extreme scenario they need to extend the oxygen supply from 30 hours to 72 hours, the outer edge of any rescue scenario.

The light is kept on for a few minutes as the men take stock of the disarray and begin to rearrange their new living quarters. The 90° adjustment means equipment previously directly in front of them is now right above them. The portholes that were once three little windows have become 'skylights' out onto the black water above. The hatch is no longer in the ceiling, but at their elbows, like a door to another room. The sonar units are at their feet, while the depth sounder, machine control gauge and switches, as well as the 1,000-watt quartz iodine exterior light, are all above their heads.

The first thing to do is reposition the two beds and foam mattresses, which they move from the vertical to horizontal. Previously there was a 12-inch gap between the bunks to allow movement, but instead they squeeze them together, like two singles into a double. The benches are now resting on the penetrators at one end and the hydraulic valves at the other. The aim is to have everything within easy reach in the dark.

There are two clockwork timers screwed to the sphere. They have a plastic dial, like an egg timer, and are used to trigger an alarm every 30 minutes to prompt the activation of the scrubber. If both men fall asleep and fail to switch it on, there's a good chance they will never wake up, as an excess of CO_2 will tip sleep into a fatal unconsciousness. The timers have a ridged, raised surface, allowing the men to 'feel' the time without switching on the light. Chapman unscrews both timers and hands one to Mallinson, keeping the other for himself. They will now be able to feel time passing through their fingertips.

The coiled lead on the microphone of the underwater telephone is long enough to dangle down in front of them, while the control box is an arm's length away. The main oxygen supply is no longer easily accessible, as it's wedged beneath Mallinson's bench, so they decide to switch to using the reserve bottle to bleed oxygen to replace the CO_2 as it's three inches from Chapman's head and the gauge is, quite literally, staring him in the face. The most vital system is now the most inconveniently positioned. The CO_2 scrubber is positioned above their feet at the other end of the sphere; this will require one of them to move around every 30 to 45 minutes, at a cost of effort and, as a result, of oxygen.

Moving around the sphere, putting in place what limited practical solutions he can, Mallinson is thinking about their plight. They are so deep and their potential rescuers so very far from the mainland that, surely, any help is impossible. He decides not to translate thoughts into words and share his fears with Chapman. Instead he lets them slowly roll around like black marbles. Mallinson is also thinking about the batteries. He can tell that the 120-volt battery has been drained down to 100 volts and is silently berating himself for spending too long on the cable burial, which has clearly taken its toll on the supply.

Chapman makes an inventory of their supplies, which are far from lavish – they have a single and very soggy cheese and chutney sandwich and a can of Corona lemonade. Mallinson reminds Chapman that he doesn't like cheese and chutney, and regrets having eaten his own strawberry jam sandwiches. There's one half flask of black coffee, a tin of powdered milk, a packet of sugar, two apples and some paper cups. There are also three soggy biscuits and the standard lifeboat ration of glucose tablets and biscuits. There's no water.

It's about now that they become aware of the incessant din inside the sphere. This has been a constant since they crash-landed, but a mixture of anxiety, adrenaline and the distraction of emergency tasks has served to render it inaudible. Now, it's all they can hear. Almost every second a loud tone reverberates through the sphere. It's the pinger on the CANTAT cable, dropped as a guide for the next shift at the exact spot where they had finished working. The repetitive piercing ping is deeply uncomfortable, but the only way to silence it is to turn down the underwater telephone (UQC), which would cut them off from all surface communication.

In this state of semi-order, Chapman reaches for the communication breaker, which suddenly bursts into life.

'*Pisces, Pisces*, this is *Voyager* … Do you read? Over.'

Chapman pushes the transducer down, which in the new disorientated world means pushing it up: 'This is *Pisces*. Yes, we read you loud and clear.'

Chapman details their position and condition. He explains that the sub is propped at a 90° angle to the horizontal, that they are uninjured and maintaining morale, and that *Pisces III*'s life-support systems are intact. He then gives as detailed as possible a reading of their oxygen supply. Yet what both men don't know is that their figures and the estimated rescue deadline given by Barrow – who have been handed control of the operation by *Voyager* – is based on an error. As the main oxygen tank is buried and therefore out of sight, the reading Chapman passes on to the surface is the last one taken, which was on the surface and seconds before the sudden sinking. The oxygen content read as 2,400 litres, but the cold fact, unknown to both men and to their rescuers, is that the actual reading is 1,900 litres – a fifth less. On the way to the seabed they have consumed the equivalent of 16 man hours of oxygen at the restricted rate of inhalation. The figures they are passing on are incorrect and any deadline Barrow sets will be out by hours.

They have less oxygen and less time than either they or their rescuers believe.

The temperature inside the sphere is 50°F (10°C), but combined with humidity running at 95 per cent it feels much colder. Mallinson is in a thick red sweater; Chapman is wearing only overalls but is beginning to feel the chill, so he strips down his overalls to his waist, puts on a lifejacket, binds his arms with a few woollen rags left over from when the sub was recently cleaned and then buttons himself back up. Both men are still shaking from a combination of cold and adrenaline.

Another concern both men have is that the batteries might fail, rendering them unable to communicate with the surface, light the sphere or operate the scrubber. The batteries are all on their side and the acid in which they sit could be draining to the bottom, thereby reducing connectivity. There could also be an unseen acid leak, but each time one of them rises from their bed to check the voltage on all three, he lies back reassured. For now.

They are three hours into their ordeal and already they recognise an issue with the scrubber. On a normal dive it takes just ten minutes to reduce the CO_2 level below 0.5 per cent, but the time taken to achieve this is rising. The scrubber has not been changed since the previous nine-hour dive and condensation is seeping into the pellets, reducing their efficacy. Chapman discusses the options with Mallinson. They have two fresh canisters but neither man is anxious to make a switch just yet. Even when they do, condensation will remain a problem. Mallinson suggests that they take turns shaking the canister while the scrubber is running in the hope that this will help, and that they let the CO_2 levels rise higher than normal to make the canister last a little longer.

In a notebook later transcribed into his pilot's log, Chapman records:

General state of health: No injuries after impact. Breathing fast because of work and initial fear of situation. Thirsty … Advice given from surface to maintain atmospheric pressure and no physical exertion.

On the surface, Ralph Henderson is scribbling down a detailed message in the radio room on *Voyager*. There's a round porthole covered by an orange nylon curtain, and whenever he looks out of the window all he can see is the grey-green expanse of the Atlantic. It's 12.47 pm and the men's morale is, he writes, 'fantastic'. The notes continue:

O2 1 bottle 3400, other bottle 2600, 2 Lth cans scrubber works. Battery volts 110. Attitud[e] 90 degrees stern down. A Tanks possibly flooded. Air bottles 2x2000 PSI 2(*1) 3300 PSI will call every ½ hour.

Henderson can also feel the frustration building. Captain Len Edwards, master of the ship, and usually a man of a preternatural calm and confidence that's surprising for a man only in his early 30s, is anxious to leave the scene and return to Cork. He knows the sooner they can leave, the sooner they can return – and the quicker Chapman and Mallinson can be recovered. But Edwards and *Voyager* can't depart the scene until they have replacement cover on the accident site, and cover has yet to be found. As he has handed over operational command to the base back in Barrow, he is now awaiting instructions. His priority at the moment is using all his skill to maintain *Voyager* over the exact spot beneath which he believes *Pisces III* will lie. The wind is freshening up and making this increasingly difficult.

For Edwards it's also proving a long, frustrating afternoon. While a number of ships have offered assistance, each is too far distant. The Irish Navy would have been his best hope, but when a Captain Kavanna called Barrow it was to explain, apologetically, that their nearest ship is 16 hours away. Edwards is now resting his hopes on the arrival of either *British Kiwi* – a commercial BP tanker – or a Royal Navy vessel, *Sir Tristram*. Edwards knows *Voyager* is lying 150 miles from Cork, which is 13½ hours' hard sailing if he can hit top speed, which is 27 hours as a round trip. When you factor in the loading time of rescue equipment and submersibles, he will be lucky to be back on this spot within 30 to 32 hours. Every hour he spends waiting is one more hour struck off his men's oxygen supply.

Henderson, meanwhile, is focusing on keeping both men on the bottom as calm as possible. He's already made a decision to restrict all bad or unpleasant news and spin any mishaps on the grounds that what they need is a sense of steady control, mixed in with a few crude jokes on how two men trapped in a sunken submarine can both keep warm and while away the hours. In the early afternoon he advises Mallinson and Chapman that if they are cold they should unzip the seat covers wrapped around the day beds and use them as an impromptu sleeping bag. Good advice, except the coverings had recently been changed; now they're fixed and so no longer unzip. The suggestion does prompt the pair to find the rubber cover that usually sits over the video equipment and use it as a blanket.

At 2.55 pm many of the crew on *Voyager* head up on deck to watch for the arrival of the RAF Nimrod. Edwards had hoped that the communications plane would be able to stay on the accident site and so relieve him and his crew, but Barrow has informed him that the RAF are refusing to accept responsibility and that *Voyager* is ordered to remain on station until a surface relief vessel arrives later in the afternoon. Upon arrival, the pale grey plane arcs overhead and circles at a radius of two miles, slowly dropping in altitude until it's around 300 feet above the surface of the sea. A bay door then opens and a sonar buoy tumbles through the air and splashes down into the Atlantic, sinking at first, before bobbing back up to the surface. Inside the Nimrod, RAF officers, whose principal task is detecting Soviet submarines, are now scanning their sonar screens for signs of *Pisces III*. Their persistent question, '*Pisces III* do you copy?' will remain unanswered.

At 5 pm Henderson tells the men to keep listening as he is about to try another transducer, but in *Pisces III* all Chapman and Mallinson can hear is the persistent 'ping' of the pinger, once every one and a half seconds. They have turned up the underwater telephone's loudspeaker to maximum and the ping is enough to induce a headache. But they can hear nothing else, and after 45 minutes of silence from the surface are beginning to be concerned.

Shortly before 6 pm a ship is spotted on the horizon. It's HMS *Sir Tristram*, a multi-purpose 6,390-ton fast troop and heavy vehicle carrier, with a crew of 18 officers, 50 ratings and, conveniently, a pair of Wessex helicopters positioned on the fore and aft decks. As the captain positions the ship 50 metres out from the buoys marking the spot where *Pisces III* sank, Henderson and diver David Mayo prepare to transfer across. A message is sent to *Pisces III* – which they do hear – that there will be a break in communications, deliberate this time, as the pair swap ships, but that they 'will be in touch again soon'. A 16-foot Gemini inflatable is prepared by *Voyager* and Henderson loads it up with a heavy 10-inch reel-to-reel tape recorder, as well as the underwater telephone system and his log books. By 6.15 pm Henderson and Mayo are on the bridge of *Sir Tristram*, being warmly welcomed by the captain and crew, and are setting up the equipment and watching through the naval vessel's vast windows. The engines of *Voyager* start up and her bow slowly begins to turn as the propellers send a flotilla of sonic waves down through the light-filled fathoms and into the darkness.

Lying in the dark of the sphere, with the underwater telephone turned up, Chapman hears a background thrum rippling underneath the repetitive din of the pinger. It takes a second or two for him to recognise the noise as the thrum becomes more consistent, picks up pace and rises into a louder thrashing churn. The sound of *Voyager*'s propellers is reaching them. It's a moment of profound emotion that neither man is willing, or perhaps even capable, of expressing to the other. Their support ship, their friends and colleagues, are, for the moment, leaving them at the bottom of the Atlantic. While each man's cool, rational intellect is aware that the thrum of engines and churn of propellers are the sonic signatures of progress, the first line in a chain of events that, God willing, should see them hauled to the surface and safety, it's very hard to think good thoughts when buried so deep and so much in the dark. The primal emotions on which each man bites down are of fear and abandonment.

Physical effort has long been an antidote to the morbidity of depression, but in *Pisces III* movement must be rationed as cautiously as air. Chapman decides to rouse himself from his ennui with a brief bout of interior redecoration. The two handheld torches (a second had been located) are being used too frequently as a means of illuminating the barometer, which records the pressure inside the sphere, so Chapman locates the screwdriver, then carefully unscrews the barometer from the wall and positions it close to the head of his bed. In its new position Chapman checks the reading: 28½ inches, a drop of two inches below the target atmosphere pressure of 30½ inches recorded on the surface. The drop is within an acceptable range and occurs whenever the scrubber is switched on, as the pressure drop corresponds with the removal of CO_2 from the atmosphere.

The problem is not with the barometer reading, but the lithium hydroxide canister. It's failing, taking ever longer to scrub the air clean, but Mallinson and Chapman are determined to push it to the limit. Chapman's other concern is their current rate of oxygen consumption, with every glance at the oxygen tank spiking his anxiety. They are still using far too much. As the barometer and oxygen tank are both now by his bunk, it's he who raises the warning.

Both men know that *Voyager* is going to be gone for a minimum of 30 hours – a span of time longer than their entire oxygen supply would last if used at the carefree rate of 1 litre/

min/man during a normal dive. They have to keep their oxygen consumption down lower, much lower than they have managed so far. They agree that as evening gives way to night they will try to remain silent and grab what brief snatches of sleep they can, in between maintaining the scrubber. Mallinson sets his timer for two hours, rolls over under the rubber cover of the VTR gear and tries to rest, in spite of the pinger's devilish lullaby. Yet he can't quite settle. A deep rumbling in his guts is pushing down into his bowels and he doesn't know how long he can hold on.

At 8.30 pm, around the time Commander Messervy and his team are taking off from Walney Island near Barrow bound for Cork, and *Voyager* is 40 miles distant from the accident site, bound for a rendezvous on the Irish mainland the following morning, a message suddenly breaks the sepulchral silence of the sphere. It's the static-filled voice of David Mayo, apologising for the delay in getting back in touch. There had been communication problems setting up on *Sir Tristram*, but they have been rectified. After collecting a quick update from Chapman, Mayo signs off.

Chapman knows the canister is canned. He has been sitting up, crouched over the bloody thing for 20 minutes, and although the scrubber is running, the level of CO_2 in the atmosphere has not dropped. The combination of moisture and excessive use has exhausted the lithium hydroxide supplies inside the can. The pair no longer have a choice. They are going to have to replace it, and that is going to come at a considerable cost of effort and oxygen. Changing a lithium hydroxide canister while submerged is rarely if ever done, as all resupplies are completed on the support vessel during pre-dive maintenance. It's a difficult task, requiring a large screwdriver and a considerable amount of torque. By now the atmosphere in the sub isn't good as they've been unable to reduce the CO_2, and both are beginning to develop headaches and signs of increased lethargy.

Mallinson strips the plastic wrapping off the new canister and moves it into position while Chapman finds a screwdriver and begins to loosen the screws. The current canister is locked tight and refuses to budge. He tries again. Then again. Then again. Over the next 15 minutes he strains to remove first the top and then – eventually – the bottom, calculating as he goes along how much longer it's taking than usual. By the time the new canister is in place, he figures he's spent three times the time and effort, all the while breathing in tiny particles of lithium hydroxide dust that are burning his throat. He winds up panting and light-headed. Mallinson prepares him a cup of cold coffee, which he shares, along with a single glucose biscuit. They switch on the scrubber and let the new canister work its magic for 30 minutes as the CO_2 level begins to drop. In his notebook Chapman records: 'Changed LiOH canister. Total used on can 60 hrs approx.'

At around 11 pm, over 13 hours since the accident and 22 hours since the sub first launched, Mallinson feels a pain in his stomach and a strain in his bowels that he can no longer resist. Both have known it was only a matter of time before this particular indignity would have to be faced, a dress rehearsal of a more convenient form of ablution having taken place earlier in the afternoon. Then they'd discovered that their 'bathroom facility' was temporarily out of order: the half-gallon detergent bottle into which the sub crews urinate was filled with cold, stale urine as it hadn't been emptied by the previous crew, and there was the added issue of the cap, which was missing, lost among the debris cluttering the floor and too expensive in oxygen expended to retrieve. (How the sub wasn't soaked in piss during the descent, neither man can even begin to explain, though both are relieved for small mercies.) The solution was to pour the half a gallon of urine into one of the black plastic bags kept on board for clearing up whatever litter routinely congregates around the place and then quickly knot the bag, but even though Mallinson had carefully stored the bag under his bench, the air is still filled with an acrid tart sting.

But now Mallinson really needs to go and it means breaking the iron rule: no one ever defecates in a miniature sub. Ever since that pie from the pub, which he can still taste four days on, he hasn't felt right. The diarrhoea of the previous day had seemed to pass, but what he does know is that he can postpone the inevitable no longer. Chapman senses from Mallinson's body language and agitation that something is bothering him. Mallinson says he's sorry, but he can't wait any longer. The fragility of the atmosphere is such that Mallinson is loath to add to the contaminants, especially as he's the butt of a running joke among the crew that his constipation is such that he requires only one trip to the toilet per voyage, an image wrecked by his recent bout of the trots. Still, he had hoped that Chapman would be the first to cross this foul Rubicon.

The upcoming procedure requires careful consideration, as the stench of faecal matter cannot be allowed to make a desperate situation any more intolerable. Together they compose a plan to minimise the odour. First Mallinson squats while awkwardly holding the plastic bag that previously wrapped the lithium hydroxide canister. Chapman turns away to afford his colleague what dignity he can, but this doesn't spare him the strain of Mallinson's grunts and the olfactory assault that marks the task's completion. Mallinson hoists up his trousers, then quickly knots the plastic bag. Chapman has already emptied the little metal first aid box of its contents and Mallinson wedges his deposit inside, presses down the air-tight lid and places it at the bottom of the sphere. Chapman knows he will be next to undergo the awkward squat; he just doesn't know when. For Mallinson any feeling of residual embarrassment is subsumed by a joyful feeling of relief, and as the scrubber is run he lies back and breaks into a smile so wide that Chapman can see the glint of his teeth in the dark.

CHAPTER SIX

Sir Leonard Redshaw is in Barrow, giving a guided tour of the shipyard to Lord Clitheroe, the Lord Lieutenant of Lancashire and the Queen's representative to the county. Lord Clitheroe is there in plumed hat and gold braid, rubbing shoulders with men in oil-stained boilersuits, to present the Vickers Shipbuilding Group with the Queen's Award to Industry on account of their export performance.

Redshaw is a tall, serious man with thinning, swept-back hair, who favours dark suits and ties, and crisp white shirts. His glasses are wireless and unostentatious, and the only colour in his outfit comes from the gold watch with its elasticated metal band on his wrist. The son of Stanley Redshaw, the Chief Naval Architect, he has risen far above his late father's role, and is now managing director of the Vickers Shipbuilding Group. Born in Barrow, he remains devoted to the town where he has spent almost all his working life. He believes in two things: 'free trade and Furness', although the unions believe the former impacts on the latter and have an intense dislike for his hard-nosed attempts to speed progress and dilute their power.

Redshaw is one of the nation's great shipbuilders, having pioneered prefabrication and the replacement of rivets with welds. The last decade has seen him oversee the construction of Britain's first nuclear-powered submarine and then her first nuclear-armed submarine. He is one of the key people who has led Britain into the nuclear age and on the shop floor he is still called 'Mr Polaris' (a model of the distinctive black-and-white missile sits on the marble mantelpiece in his office). Formidable in a dispute, he will go eye to eye with any admiral who doubts his, or his men's, ability to deliver.

He also likes to remain in control. Whenever they approach the Kirkstone Pass, the tight, steep and winding road linking Windermere in the south to Ullswater in the north, Bill Moorby, his official chauffeur, is made to sit in the passenger seat as 'the boss' takes the wheel of the company's powder-blue Jaguar. His biographer noted, 'Redshaw squared up to hazards with confidence,' and, as his chauffeur could attest, at speed.

When Redshaw is informed of the accident shortly after 10 am on Wednesday morning, he breaks off his tour and immediately returns to his office. The plight of *Pisces III* is personal, triggering memories of old black-and-white newsreel footage and photographs from over 30 years back when HMS *Thetis*, a Royal Navy submarine, sank in Liverpool Bay. The *Thetis* had been on diving trials in 1939 with 103 men on board, twice her normal complement, as technicians crammed on board to monitor the vessel's performance. The inner door of a torpedo tube was accidentally opened while the outer hatch was inexplicably open to the sea. The sudden inrush of water sank the bow of the sub 150 feet to the bottom of the seabed, while the stern and propellers were raised above the surface.

A number of delays followed, first in registering that the

submarine was stricken, then finding her exact location, so an attempt at evacuation did not begin until 20 hours after the accident. The escape pod on *Thetis* was single-man: a submariner was required to hold his nerve in a sealed chamber, which then flooded with water, equalising the pressure inside to that outside and so allowing the opening of the hatch. Three men successfully escaped, but the fourth panicked, attempted an early escape, then drowned, in the process jamming the hatch and preventing further escape. The remaining 99 men died of carbon dioxide poisoning, including four men from Vickers. The official inquiry, which Redshaw had read, revealed a long list of human errors, slow communication between different departments and, most fatally, a failure to swiftly establish a centralised command around an experienced, competent figure.

In the time it takes Redshaw to walk to his office, he has grasped the nettle. He is taking control – the buck will stop with him and the two men's lives are now in his hands. Their deaths will be his fault, and no other.

Greg Mott had been in charge of the special progress department during the construction of HMS *Dreadnought*, Britain's first nuclear-powered submarine, and had used colour pegs plotted on a wall chart for the monitoring of the daily progress of every item of equipment to help predict bottlenecks and prepare timely solutions. He is a man Redshaw wants by his side but he's in the London office at the Vickers Tower when he gets the call. He immediately arranges to travel back to Barrow.

* * *

At the headquarters of Vickers Oceanics, a ramshackle cluster of Portakabins formed around the 'loco shed', where the *Pisces* crafts are wheeled in for repairs and refits, Commander Peter Messervy is marshalling his men. The call from *Voyager* comes in by radio telephone from Ralph Henderson shortly after 10 am. A meeting is interrupted as Messervy, the technical manager Harold Pass and everyone else present gathers around the speakerphone. The report is succinct: P.III Aft Sphere flooded, with the submersible now bottomed at 1,575 feet. (A telex will later come in, with a transmission error, incorrectly stating the depth as 1,375 feet, the second digit wrongly transposed from a '5' to a '3'. This will be the figure widely quoted to the press and, for a time, among the rescue team.)

A telex is also received from the telegram agency Subkable London to Vickers Oceanics, Barrow. It reads:

Attention Commander Messervy stop
Account of Pisces Three Accident stop
Normal recovery proceeding in Force 8 wind stop.
When diver had made fast tow rope to Pisces considerable slack rope seen stop
Bight of this slack washed over Pisces by wave action stop
Prints arriving Barrow 2203 Tonight Red Star stop.

Messervy knows immediately that the only way to get the sub and the men back to the surface is if another submarine can attach a rescue line. No diver could reach such a depth. Almost three years earlier a safety diver named Sheck Exley hit 465 feet just off Andros Island in the Bahamas when he went in search of a missing dive team. He almost died, blacking out from narcosis, and that wasn't even a third of the depth at which *Pisces* is trapped. Messervy reviews the equipment to hand and starts to make decisions. He only has to look out the window to see *Pisces I*; she's in the loco shed for an extensive refit and isn't going to be available any time soon (though the sub will be 'cannibalised' for parts, including the removal of the manipulator arm as a spare). *Pisces II* is on board the Vickers *Venturer* working in the North Sea, currently 150 miles from Teesport, while *Pisces IV* and *Pisces V* are in Canada, the latter 2,500 miles from Cork on Canada's east coast, the former 4,500 miles away on its west coast.

Three minutes after the initial call from *Voyager*, Vickers Traffic, the logistics department, is contacted and told to start figuring out the fastest and most efficient way to get *Pisces II* to Cork, the closest port to the accident site. Two minutes after this call, at 10.05, a priority radio telephone call is booked by the switchboard to go from Messervy's office to *Venturer* in the North Sea. Calls offshore take time, and Messervy knows he may have to wait up to an hour.

At 10.15 the most crucial call is made, one which, arguably, should have been the first. Messervy contacts Her Majesty's Naval Base, Clyde, at Faslane in Scotland. Based on the Gare Loch, the salt-water loch that offers quick access to the deep waters of the Atlantic, HMNB Clyde is home to the nation's nuclear deterrent, the fleet of nuclear-powered and nuclear-armed submarines. It's also the base of operations for all submarine rescues. Given their work on the construction of these submarines, Vickers are in constant close contact, while Vickers Oceanics have made it a feature of pre-planning for any eventual accident to notify the Royal Navy at the beginning of every dive and upon its safe recovery. Messervy briefs the duty officer and requests that the Royal Navy immediately initiate SUBSMASH. This is a message sent to every single ship in the fleet, wherever they are in the world, with an alert to the situation and the order to offer all necessary assistance if so requested.

At 10.35 Messervy finally gets his call through to *Venturer*. The captain explains that they are working on the Ekofisk line in the southern section of the Norwegian oil field and are at least 16 hours' hard sailing from Aberdeen. Messervy's message is clear: immediately stop what you are doing, turn around and head at full speed for the nearest port.

When Redshaw and Messervy talk, the managing director makes it clear that the rescue operation has to involve the mobilisation of every available asset, preferably in duplicate, on the basis that failures and faults will undoubtedly occur but will not – repeat, will not – dictate the final result. He is clear and commanding: they will get these men back. This will not be another *Thetis* – not on his watch.

Messervy knows they need to get the Canadians at Hyco on board. At 10.46 am a transatlantic call is made to Dick Oldaker, president of Hyco. Oldaker explains that the best person to speak to is Jim McFarlane, the company's operations manager. The problem is that McFarlane is at home, in bed in Port Coquitlam, a small city 20 miles east of Vancouver, and he's not answering the phone. Oldaker figures McFarlane must be a heavy sleeper and so he calls the nearest police station, tells the duty sergeant what is going on in the middle of the Atlantic, and shortly afterwards two Canadian Mounties are dispatched to pound on McFarlane's front door. When he finally stumbles out of bed, it's around three o'clock in the morning, and he's surprised to see two uniformed police officers who tell him to answer the phone.

McFarlane knows Hyco are in a good position to support any rescue and recovery. *Pisces V*, which is then less than six months old, is already in Halifax with a full crew, preparing to work on the burial of the Canadian end of the CANTAT-2 telecommunications cable. McFarlane phones Halifax, where it's already after dawn, and speaks to Mike Macdonald, who briefs Bob Starr, Bob Holland, Al Witcombe, Steve Johnson and Jim McBeth. They've not yet been told they're definitely going – Vickers are, for the moment, examining all their options – but to start preparing in case they're needed.

Yet no sooner has McFarlane hung up after briefing Macdonald than his home phone rings again and it's Messervy, who tells him Hyco's support would be 'appreciated'. McFarlane has good relations with the Canadian armed forces, and so contacts the flag officer in Halifax with a big request. McFarlane is going to need a Hercules aircraft. The flag officer says the nearest available Hercules is in Trenton, Ontario, and that McFarlane should call and explain the crisis – and that he will too. McFarlane is a little slower on the dial, and by the time he gets through to the Air Defence Tactical Operations Centre (ATOC) at Trenton the officer says he's already spoken to Halifax, understands the nature of the emergency and that the air crew have already began their pre-flight inspection. Before he hangs up, the officer jokes that they'll worry about confirming the story and sorting out the necessary paperwork later, but that McFarlane's Hercules will be thundering down the runway within the hour.

Down at the docks in Halifax, Macdonald is hard at work with the rest of the crew, whom he has roused from assorted hotel rooms across the city. They are now stripping all the equipment they might need out of their ship, the *Hudson Explorer*, and laying it out along the quayside. He's been told that the Canadian Navy is sending a fleet of flatbed trucks, but they've not yet arrived. The wharf is filling up with battery chargers, air compressors, oxygen tanks and generators. Whenever a crew member asks whether they should take this or that piece of kit Macdonald replies in the affirmative. They are applying the old adage, 'Better to have it and not need it, than need it and not have it.' As he later recalled, 'We took everything.' Macdonald also has one specific item to source. He has to find 6,000 feet of polypropylene '2-in-1' high-strength rope. And the clock is ticking. They want to be in the air by noon.

When the naval trucks arrive, the Hyco team carefully lash up *Pisces V*, then use the *Explorer*'s miniature crane to lift her off the ship, swing her over the quayside and down onto the flat bed of the truck.

On the other side of the country, McFarlane is getting annoyed. Dick Oldaker keeps calling him for updates. McFarlane doesn't want to tie up the telephone line unnecessarily until he knows that the Hercules with *Pisces V* and crew have taken off from Trenton so he tells Oldaker, politely, to 'fuck off'.

The Royal Canadian Air Force Base Shearwater lies on the eastern shore of Halifax harbour, about four and a half miles outside the town of Shearwater, Nova Scotia. At around 11 am Eastern Standard Time the fleet of trucks from the harbour speeds through the base's front gates and straight onto the runway, where the tail flap of the Hercules is already down and awaiting the cargo. By 12 noon the plane is thundering down the runway at 90 miles per hour, then takes off. On board, Mike Macdonald can't find a seat so decides to bed down on the last thing he managed to find: the 6,000 feet of polypropylene '2-in-1' high-strength rope.

Messervy picks up his papers, fixes his monocle into his right eye, then strides out of the Portakabin and across the courtyard to the loco shed, where a group of about 20 men, the core team of *Vickers Oceanics*, are gathered around the disassembled hulk of *Pisces I*. The shed has railway tracks and once served as a repair shed for locomotives. There are overhead cranes, service pits and a sultry topless brunette Page 3 girl, cut out from *The Sun*, who gazes down onto the tool-strewn work bench. As Messervy arrives, someone reaches for the radio and Tony Blackburn's interminable jovial banter on Radio 1's morning show is silenced mid-sentence. The men are aware that something has gone wrong with *Pisces III* but do not yet know the full story. They gather around Messervy, who cannot help but feel a frisson of martial excitement. Once again lives are on the line and there are eager young men to lead.

He begins by setting out what they know: the aft sphere of *Pisces III* is flooded, causing her to sink and crash-land stern first on the seabed of the Atlantic Ocean at a depth of 1,575 feet. They have been told that the position is marked by two buoys. Mallinson and Chapman are alive, in communication with *Voyager* and in 'good spirits'. Their plight, though, is precarious. They are at a depth far deeper than any previous submarine rescue, and, as he explains, it now falls on the men in this room to figure out how to get them to the surface. Messervy also needs to know what else they should be worrying about and preparing for. Oxygen is the obvious concern, and they are working on the calculations just now, but what else?

Geoff Hall, a young electrical engineer, is concerned about the batteries. Mallinson and Chapman may have oxygen, but if their batteries fail they will have nothing to drive the scrubber and will be long dead from carbon dioxide poisoning before the oxygen runs out. He explains that the batteries are housed in GRP boxes. The battery boxes are filled with hydraulic oil. Attached to the box is a bladder of oil, and as the sub goes down, the pressure exerted on the bladder is transferred to the oil in the box, giving it a zero pressure differential. The box itself isn't pressure resistant, he stresses, and isn't designed to withstand the crushing weight of the deep. The angle at which *Pisces III* is positioned means it's likely that the lead–acid batteries will lose their electrolyte. The batteries are made by the manufacturer Oldham, based at Denton in Manchester – advertising slogan: 'I told 'em, Oldham!' – but when Hall finally gets through to the company's technical manager, 'a charming old man called Percy', his honest answer is that he has no idea if the batteries will survive.

The task of figuring out how to lift *Pisces III* off the seabed lands on the shoulders of Doug Huntington, a young 23-year-old submersible engineer, under the guidance of Ted Carter, the assistant technical manager. A few minutes before the Tannoy summoned everyone to the shed, Huntington was at his desk producing technical drawings for the ongoing refit of *Pisces I*. By the time he returns to his desk he has the weight of two men's lives and a 12-ton submersible on his shoulders. How exactly is he going to lift that load? Together with Carter, a naval architect with a background in ship design, he begins to calculate the strain that any style of hook or catching device would have to tolerate. Carter's view is that it has to be simple, not over-designed and liable to break mid-lift. They are as yet unsure if *Pisces III*'s main lifting hook is accessible to any rescue sub, but know, given the description provided by Mallinson and Chapman of the position in which they have landed, that the aft sphere should be accessible.

So what if the cause of the sinking could be the solution to them surfacing? The aft sphere is a much bigger target, with a diameter of 60 inches, compared with the 12-inch gap of the main lift hook. They need something that can be pushed into the aft sphere and, once inside, lock hard. Something like a rawlplug or an upside-down umbrella.

Huntington begins drawing a few initial sketches in pencil. The device resembles an upside-down letter 'T', the base of which contained a 'D' cuff to which the rope would be attached. The two arms of the 'T' will be much wider than the diameter of the aft sphere but bolted on and hinged, allowing them to push up and close when entering the sphere, fall back open once inside and then lock tight under the sphere's steel frame. Carter likes the simplicity and strength of the design, so Huntington quickly drafts a more detailed technical drawing to take to the fabrication shop.

The drawings, once complete, are quickly rolled up, secured with an elastic band and carried by Huntington out to the car park. He climbs into the Oceanics staff car, a white Hillman Hunter, and drives the mile or so down to the Vickers engineering works, where he asks to speak to the foreman of the fabrication shop. When the foreman arrives, Huntington rolls out the drawings and explains that he needs the device as quickly as possible. The foreman looks them over and says he can do it next week. Huntington said it needs to be today. The foreman shakes his head and says he can't do it, at which point Huntington asks to use his phone. He calls the switchboard and requests to be put through to the office of Sir Leonard Redshaw. At this point the foreman's ears prick up. Huntington tells Redshaw his idea for the device, then Redshaw asks to speak to the foreman. At the end of the brief call the foreman says it will be ready before 5 pm.

Huntington stays in the fabrication shop to keep the pressure on. While there, he discovers that Hyco in Vancouver have placed an order with Vickers Engineering to manufacture an aft sphere for a new submarine, *Pisces VI*. This means that when the design has been built he can test that it fits.

* * *

Bob Hanley is in the bath when he gets the call. He is living on Walney Island, a 15-minute drive from the Barrow office in his metallic grey Humber Sceptre. When he reaches the phone and picks it up, it's Mike Graham, the personnel manager at Vickers Oceanics. Hanley drips onto the shag pile carpet as Graham, who is audibly panicked, says that he needs to come back to the office right away.

Hanley asks what the problem is.

Graham tells him that *Pisces III* has gone down.

When Hanley reaches the office he speaks to Elliot Sinclair, who briefs him on all the details that are currently known. Later, when Sinclair is on the phone to Ralph Henderson on *Voyager*, he hands the phone briefly to Hanley. Henderson and Hanley do the same job – operation controller, or field officer, as it's known – but on rotation. During the call, Henderson, who is quite calm, says he needs all the help he can get.

Hanley heads over to the engine shed to check on progress, then drives back home to pick up some gear. The plan is to meet Messervy and the rest of the team at the airfield later in the evening.

When the phone rings on June Chapman's desk she answers it, judges from the tone of Greg Mott's voice that the request is urgent and quickly transfers the call to her boss, George Henson. The calm countenance with which Henson usually responds to calls immediately shatters, and June is straightaway aware that something is wrong. She thinks it unusual that Henson turns away from her, hangs up and then heads to the other end of the office to speak with a colleague. The two men move away from the desks so that they can talk in private, then Henson strides back across the office, collects his coat and says that they are going down to Oceanics. 'You're coming with us,' he tells June

The look on Henson's face has made June a little anxious, a feeling that begins to quicken and roll as he moves across the office, until by the time they're in the outdoor corridor it has formed into a tight, black ball of dread. She knows that something is seriously wrong and that her husband has to be in some way involved.

In the corridor, out of earshot of the others, June asks if there's a 'drama'. Henson is direct. There's a major problem at Vickers Oceanics. A sub is trapped at the bottom of the Atlantic and Roger is on board. That's all he knows, and they'll find out more when they get over to Oceanics. As they step into the lift, June can't help but replay the many conversations she and Roger have had about the fate of sunken submarines, and how, in his bitter belief, rescues were invariably futile.

The doors close and the lift begins to drop.

It's 12 noon when Messervy makes contact with Commander Ramos, the US Navy's liaison officer at United States Naval Forces Europe (CINCUSNAVEUR) in London, who offers 'all assistance'. The USS *Aeolus* is the nearest US naval vessel to the accident site and a message will be dispatched for her to make haste. Ramos explains that he intends to see what other facilities he can put at Oceanics' disposal, and will call back with further details as and when he secures support. Within hours Earl Lawrence, salvage master for the US Navy

Department, and Commander Bob Moss, the US Navy's deputy supervisor of salvage, make plans to fly in from Washington, DC.

At 12.08 Oceanics' command centre gets word back from the Ministry of Defence in Plymouth that HMS *Hecate* will be deployed with an extra lifting rope on board. *Hecate* is expected to be assisted by Force 5 or 6 westerly winds and is estimated to be on the accident site by 11 pm.

But there's a problem. For the rescue operation to begin, *Voyager* must be able to leave the accident site and sail back to Cork harbour to pick up *Pisces II*, incoming from the North Sea, and *Pisces V* from Canada. Even at top speed and with favourable winds, it's a 12–13-hour hard sail from the accident site back to the south coast of Ireland. Then factor in a few hours to load up both crafts and crew, add on the return journey back to the site, and a minimum of 30 hours will have passed – and vital air consumed – before a rescue sub can sink below the surface.

The reason *Voyager* is unable to leave is that an on-site nautical presence is vital. *Pisces III* sank after her location tether and buoy were unclipped, meaning there's no direct physical link between the surface and the sunken sub's current location. There are two plastic identification buoys bobbing on the surface, but these were positioned after the sinking and represent – at best – an estimation of the point at which *Pisces III* left the surface, giving little or no indication of the precise location where she crashed down 1,575 feet below. For now there's audio communication between *Pisces III* and *Voyager*, but as radio waves operate best vertically and become contorted and diffuse as they radiate horizontally, it's best maintained by a craft directly above the sub.

There's another possible solution. At 1.25 pm Messervy contacts the RAF Maritime Command and requests that a communications plane be dispatched to the site in order to relieve *Voyager*. The RAF agree to send up a Nimrod, which, 60 minutes later, is speeding to the accident site at 400 miles per hour. It's estimated to arrive at 3.54 pm, with an extra hour factored in to locate *Voyager*, but as we've seen, the RAF refuse to accept responsibility for maintaining contact with *Pisces III* in the event that *Voyager* leaves the site. Without such a guarantee, *Voyager* has no choice but to remain in place and wait for the first relief vessel, which is still hours out.

The disappointment at the RAF's refusal is softened slightly by the news, at 3.07 pm, that the United States Navy Salvage Department are offering the services of CURV-III, a cable-controlled underwater recovery vehicle. The offer is accepted and plans made for its arrival at Cork, where James Scott, who runs a ship-handling agency, has provided his offices on the second floor of a small office block near the port as an Irish base of operations. Scott and his team are already looking into sourcing the articulated lorries needed to get *Pisces II* and *Pisces V* the eight miles from Cork airport down to the harbour.

In addition, *Voyager* had been advised by the Ministry of Defence Naval to issue a 'Mayday' to attract the attention of any vessel, of whatever size, that could come to her relief.

It takes until 1.50 pm before Messervy gets the correct telephone number for the Irish Navy. The phone is promptly answered, and by 2 o'clock the Irish Navy say they have a ship in Dingle Bay that could be the earliest on site. And as we saw, there are two other ships who hear the distress call and change course to provide assistance: the BP tanker *British Kiwi* and the RFA *Sir Tristram*.

When June steps into the Portakabin along with George Henson, she is quickly comforted by the calm but determined activity in front of her. Peter Messervy is sitting in a chair at the centre of the newly constructed operations room and appears to be bending the entire world to his will. The rescue operation is in its initial gestation period but already growing strong arms and legs. He explains that SUBSMASH has been initiated, that the Canadians are on board, and that he will shortly be speaking to the Americans to explore what resources they've got and are prepared to make available.

Messervy makes it very clear that her husband is safe and well, as is Mallinson, and that he has no doubt that the rescue will be successful. He also explains that it's going to take time, and, while the men have a finite amount of oxygen, he knows that they can get them to the surface before it runs out.

The visit is tremendously heartening. June now has faith that her husband will be proven wrong. A submarine sunk at depth *can* be recovered. It's a fragile belief that will be tested to breaking point, but one to which she will now cling.

Just before 3 pm Al Trice rings from his bedroom at the Park Lane Hotel in London. He has kept in touch with Messervy over the past couple of years and is, fortunately, in Britain for business meetings with P&O, the British shipping company. He's already been briefed by the office in Canada, and wants to know if welding and battery-charging equipment are available on board *Voyager* or whether his team should add these items to their list. He also flags up the slightly inconvenient fact that two of Hyco's pilots will be flying on board the Canadian Air Force Hercules without their passports and somebody needs to contact Irish border control to smooth their arrival. There can't be delays, filling forms or interviews with Irish jobsworths. Trice confirms that the Hyco team are flying over floating line, hook 27kHz array and underwater telephones.

He tells Messervy that he expects to arrive in Cork by 10 am the following day and will either be at the airport or at the Rank Hotel in the city centre. What he doesn't know is that his wife has other plans. The couple have travelled to England with their young daughter on her first trip to Britain, and Al is going to have to get both his wife and daughter settled with friends in Oxford before he can go anywhere.

Communication between *Venturer* and Base Ops Team Barrow is breaking up. The team at base unfortunately can't just pick up the phone and speak to the captain on the bridge and get a clear, eyewitness view. They are using a combination of longwave and shortwave radio telephones, and static is eating into every conversation, chewing off chunks of words and phrases. Redshaw points out that they do have one of the most sophisticated pieces of engineering on the planet laid up in the dry dock, HMS *Swiftsure*. If a nuclear submarine can communicate with Britain from under a polar ice cap, surely it can reach a vessel 150 miles off the coast of Ireland. In time the frustrating answer will come back: it certainly could reach *Venturer* if the sub were under a polar ice cap, but in Barrow's sheltered bay the surrounding hills are acting as a barrier.

It's 5 pm in the North Sea, and the oil rig supply vessel *Comet*, on charter to Phillips Petroleum, is manoeuvring alongside the Vickers *Venturer*. Bob Eastaugh, Oceanics' lead operations manager who has been co-ordinating the North Sea dives with *Pisces II* from *Venturer*, knows that their vessel is too slow. So at lunchtime he and the captain began calling out for assistance, seeking a faster ship, but one that can also carry a heavy load. The *Comet* called back to say she could carry *Pisces II* on her deck and could go four knots faster, shaving three hours off the time it would take *Venturer* to reach Teesport docks. A rendezvous point was fixed.

Once alongside, *Comet* slows to a stop, then carefully manoeuvres the arm of its crane over the side and down onto the deck of *Venturer*. Eastaugh and his team, which includes the company's chief pilot Des D'Arcy, have already got *Pisces II* in her cradle, and it doesn't take long to hook her up. The crane operator gently lifts the sub up, swinging her over and down onto the deck of the *Comet*, to which she is securely lashed.

Eastaugh, D'Arcy and the rest of the team then leap over the deck rail onto the *Comet*, and within a few minutes they're waving *Venturer* off as they – and *Pisces II* – pick up speed and head south. The hope is that they can make Teesdock by 9 pm, or 10 pm at the latest.

* * *

A light dusk has begun to settle on Walney Island, where Peter Messervy is finally taking his seat in the company's blue-and-white six-seater single-engine Piper Comanche. It's 8.20 pm.

He is joined by Nigel Hook from Vickers Oceanics, who has just been appointed OIC (Officer in Command) Cork, Bob Hanley and Doug Huntington, who arrives laden with two of the freshly minted heavy steel hooks or toggles – no one can quite decide how best to describe them, other than exceedingly heavy. It takes two men to get them out of his car and into the hold. The pilot is concerned lest they tip over their prescribed payload weight, but figures they should just about be OK.

Messervy is sitting beside the pilot in the cockpit, clutching his black executive-style briefcase on his lap, with the paperwork he will read on the flight, including the most accurate estimate of the oxygen levels within *Pisces III*. According to the report they have until 8 o'clock on Saturday morning to get the submarine and the men to the surface with the hatch open to the air. But for now, for a few seconds, he relaxes, confident that all the parts of the rescue plan are now in play. All the parts, that is, but one – the Americans.

Larry Brady is woken at 4.45 am at his lakeside home, a 30-minute drive from San Diego. A tall man with a salt-and-pepper beard and shaggy, shoulder-length hair, he reaches for his wire-rim glasses sitting on the bedside table.

The call is from his boss Mason Wright on the *YFNX-30*, CURV-III's support boat, anchored down at Point Loma in San Diego Bay. The message is direct – 'Larry, we just received a "sub lost" flash message.' Wright explains that a small sub is lost off the coast of Ireland and tells Brady to mobilise the CURV, of which Brady, a confident 37-year-old, is the principal pilot. Before Brady hangs up, he asks Wright to pull out the nuts and bolts that secure the soft top over *YFNX-30*. The cover needs to be off before they can remove their equipment. Wright tells him he's already on it. Brady rises, dresses quickly and begins calling the rest of the team.

Bob Watts, CURV-III's programme manager, is already briefed, but Brady calls up sonar technician Tom Wojewski, deck supervisor Denny Holstein, mechanical technician and test engineer John De Friest, electronics technician William Sanderson and photo technician William Patterson. To each one he says the same thing: 'No one calls off sick. This is an important day.' Everyone is on board.

If there's a still calm centre of the universe for submarines in 1973, it's San Diego. The US naval base is just across the water, and in the surrounding neighbourhoods is an array of marine support companies specialising in optics and cables. The ocean technology division of the Naval Undersea Center is headed by Howard R. Talkington and based on military property at Point Loma. Home is an office block that backs up onto a couple of piers where the operational barge *YFNX-30*, a long, low-lying steel-hulled vessel, 110 feet long with a 34-foot beam and diesel engines, is moored.

The Naval Undersea Center isn't military – there are no snappy salutes required – but the US Navy is the main client. It's divided into five different divisions, with 20 staff currently at work in the ocean technology division, which is pioneering remote-operated vehicles and shallow-water inspection vehicles. The US Navy also has *Mystic*, the DSRV-1 (deep-submergence rescue vehicle) designed by Lockheed at a cost of $41 million and launched three years ago in 1970 for rescuing crew from sunken nuclear submarines. Equipped with two pilots and two rescue personnel, and able to operate at a depth of 5,000 feet, it can carry 23 men at a time. The navy said they would be happy to deploy, but it would take 48 hours and there are questions about the compatibility of *Pisces III*'s hatch.

Remote underwater vehicles were developed in the late 1950s and early 1960s when the Navy Bureau of Ships put out a call to American industry to develop a manoeuvrable underwater camera system. The navy wanted to develop a more efficient means of retrieving lost ordnance, and by 1961 they had XN-3, which was remotely operated and featured both a camera and a clamshell claw. Over the next five years XN-3 evolved in both design and depth capability into CURV-I, which debuted before the eyes of the world in 1966, when 31,000 feet above the coast of Spain a B-52 Superfortress returning from a mission over Soviet airspace and carrying four Mk28 hydrogen bombs collided with a KC-135 flying tanker during mid-air refuelling. The flying tanker exploded after the fuel ignited, killing all four crew members. The B-52 also broke apart, but four of the crew survived.

Three of the four hydrogen bombs landed on the outskirts of a small fishing village, Palomares in the Almería region of southern Spain, and were quickly recovered, though the conventional explosives in two of them had exploded, contaminating a two-kilometre area with plutonium. The fourth hydrogen bomb was spotted landing in the bay by a fisherman, Francisco Simó Orts, forever after known locally as 'Paco el de la bomba' or 'Bomb Frankie'. The search to locate the sunken hydrogen bomb took ten weeks and was finally resolved when CURV-I was able to attach a lift line to the bomb. CURV-I did become entangled in the bomb's parachute, which meant it was unable to retreat from the bomb after securing the line, so both bomb and miniature sub were raised together. Seven years on, and CURV-I had evolved into a third iteration, with greater manoeuvrability, aquatic agility and control.

Shortly after 6 am, Larry Brady and his team are dockside, with dawn's rosy light colouring the water as they are hard at work on *YFNX-30*, preparing CURV-III for its transatlantic flight. The barge's soft roof has been removed and the team are working on the support equipment inventory: cable, lift lines, floats, control van, crane, supply van, spares, motor generator, capstans. The cables that connect CURV-III to the surface are folded within containers, eight feet long and five feet high, that sit like steel sarcophagi on the deck.

CURV-III has an aluminium frame on which blocks of syntactic foam are balanced to provide a degree of positive buoyancy. It's seven feet high, five feet wide and 13 feet long, weighs 5,000 lbs and is tethered by a 10,000-foot-long multi-conductor cable that runs up to the surface, where it connects to a control centre secured inside what the Americans describe as a 'small van', but which looks like a large shipping container. When operational, CURV-III is equipped with a still camera, lights, two TV cameras, active and passive sonar altimeter, and a depthometer. The TV cameras can both pan and tilt, and their field of vision is illuminated by four 250-watt headlights and one mercury spotlight on each pan and tilt unit. The manipulator arm can carry up to 200 lbs.

Brady and the rest of the team think they have everything packed, even a couple of bags of candied orange slices and jelly beans, courtesy of the wife of Bill McLean, one of their colleagues. Once loaded, they start up the engine of the barge and head into the sunlit waters of San Diego Bay and across to the Naval Air Station on North Island, where two C-141 Starlifter aircraft are waiting for them.

Unfortunately, CURV-III and its equipment are not easily compatible with the holds of the aircraft. The old-style lifting equipment proves problematic and it takes longer than expected. As Brady recalls, '[with] a Rube Goldberg array of forklifts, pallets, jury-rigging, elbow grease, Vaseline and the body language of Talkington, McLean, Capt. Gautier and Schlosser, we managed to cram the material into the two planes'.

Afterwards, Brady has just enough time to quickly drive home, pack a bag of clothes and return to see the final addition to the aircraft: a pallet of temporary aircraft seats on which the team will sit in the cargo hold. They all strap themselves in and prepare for take-off at 7.30 pm, Pacific Daylight Time. There's a temporary stopover in Delaware, to refuel and fix a repair to the black box on one of the C-141s. The team are also passed an array of packed lunches to sustain them on the flight to Cork.

* * *

Back in Britain, the *Comet*, the oil rig supply vessel carrying *Pisces II*, hits choppy waters that reduce its speed and delay its arrival into Teesdock by almost two hours. It's 11.45 pm when Bob Eastaugh and his crew begin to offload the submarine. A 25-ton and 22-ton crane are on standby, and the sub swings gently down onto the back of an articulated lorry, then travels under police escort to Teesside airport, where an RAF Hercules is waiting, fuelled and ready, on the floodlit runway. Take-offs after 10 pm are prohibited at this regional airport, but this will be an understandable exception. It will be 2.30 on Thursday morning before the Hercules's wheels are up, with an estimated time of arrival in Cork of 3.55 am.

In the second-floor office by the waterfront of Cork harbour, James Scott and his team are preparing to work through the night. He now has the estimated arrival time for both the Canadian Hercules carrying *Pisces V* and the RAF Hercules carrying *Pisces II*. Transport trucks are in position at the airport, and he has the tugs, berth and cranage available at the docks from eight o'clock in the morning, a few hours before *Voyager* is expected to arrive. What he doesn't yet have in hand is a coaster with a 10-ton lift.

PART III

THURSDAY

CHAPTER SEVEN

On the surface, David Mayo is accidentally causing a royal kind of confusion. Mayo is out on the deck of the *Sir Tristram* in the dark, hunched over the underwater telephone, whose transmitter hangs over the side and is submerged a dozen feet down. Among the vessels that offer support on Wednesday is the Cunard transatlantic liner the *Queen Elizabeth 2*, launched six years earlier at John Brown Shipyard in Clydebank. The offer of assistance is appreciated though unnecessary, but her telex of support is about to bring a surprising buoyancy of spirits to Chapman and Mallinson. Communications have been patchy since 8 pm on Wednesday evening, and it's now 1.30 am when their underwater telephone rings. The static is loud and the volume low, but the message both men can hear is: 'Best wishes to *Pisces* crew and hope all goes well, from Queen Elizabeth.'

The submarine is briefly transformed by the heartfelt benediction only a monarch can bestow upon a loyal subject. For Chapman, who for years had vowed as a Royal Navy submariner to serve Queen and country, the idea of Queen Elizabeth II being aware of their predicament and finding herself so affected as to send a message of support is deeply moving.

Mallinson is impressed by the Queen's thoughtfulness and feels an actual warmth rise up from his feet. For a few minutes at least, the dark loneliness of their position is banished, replaced instead by the brilliance of Britannia.

The first night is tough. The temperature reads 50°F (10°C), but the humidity is making it feel much colder and damper. The curved walls of the sphere are dripping with condensation, the pair's every breath transformed into vapour, adding to the chill and raising their levels of anxiety. When a droplet builds to become a drip and lands on Chapman's forehead, his first thought is that the outside water pressure – 800 psi, or the equivalent of 50 tons on the hatch alone – has found a faulty seam, a weak bolt, a loosened joint, and that this initial leak is but a harbinger of the sub's imminent collapse. So he immediately switches on his torch and begins to scan the sphere for signs of a leak, while simultaneously tasting the water that has splashed on his face. Is it salty?

It isn't. Chapman relaxes. For now.

To increase body temperature and combat humidity's chill, both men are 'cuddling up', to use Mallinson's phrase, lying side by side, shivering under the plastic cover and occasionally rolling over to 'spoon'. *Morecambe and Wise*, the popular BBC1 show in which the comedy duo share a double bed dressed in striped pyjamas, could be an appropriate reference; what would those two do if stranded 1,575 feet below the Atlantic Ocean? It's hard to imagine them each clutching a timer like Chapman and Mallinson. Sleep does come but it's thin, a dance of disturbed dreams, and when the alarm goes off and Chapman is roused, the first few minutes are full of befuddled confusion as he struggles to remember where he is.

And then realises. 'Those,' he recalled, 'were the worst moments.'

Over the course of the night, Chapman forgets to check the oxygen level at least twice. Mallinson wakes up at the same time as Chapman but, as he's not responsible for activating the scrubber, he rolls over and tries to sleep, but instead worries about his wife and children. If anything happens to him, if the two of them don't survive, how will Pamela cope raising three children on her own? He knows Chapman is just recently married, and that he and June have yet to start a family. He doesn't know if this is now a comfort to Chapman or a source of regret, but like the Englishman he is, he thinks it best not to enquire. Emotions, like blood and water, can go everywhere when spilled.

As a source of distraction Mallinson begins to play organ music in his head, sometimes moving his fingers as if to mimic a keyboard. Other thoughts, less distractions than condemnations, bubble up when he tries to figure out how they got into this situation. He's sure it has something to do with the repairs he requested on the aft sphere – repairs that weren't carried out. He doesn't yet know that the entire aft sphere hatch has ripped off, but he's aware that something gave way to allow the water to pour in at such a pace.

Then there's the possible means of their rescue. Mallinson lies in the dark, thinking about the strongest place to fix a lift line. He thinks about the submarine's blueprints, turning them by 90° in his mind to more accurately resemble their current position. When he tries to talk to Chapman about his theory that the forward end of the frame is best capable of withstanding the structural stress during the ascent, he can sense that his partner is unwilling to fully engage. At first he thinks Chapman might doubt the chance of a rescue, but he soon realises he's less concerned about a rescue attempt over which they have zero control than eking out the oxygen supply so that it will last until they're rescued.

Chapman also has a different, more trusting, attitude towards Messervy and the team, thinking that they can be relied upon to figure out the best approach. Mallinson, meanwhile, is sceptical of anything done by anyone other than himself. He also likes to think aloud, which Chapman considers a waste of air, so when his partner's wheels start to spin, he refuses to answer or join in the speculation and instead hopes that Mallinson will take this not-so-subtle hint. As Chapman recalled: 'Our last instruction was to act as vegetables most of the time, but still perform as intelligent human beings every half-hour for essential life support.'

The ability of humans to control their breathing is limited but can still be effective in certain circumstances. On average, a person will breathe in and breathe out 20,000 times over a 24-hour period, using 12,500 litres of air, an amount that will expand or contract depending on the size of the person and their level of exertion. The molecules of oxygen in the air breathed through the nostrils then pass through the sinus cavity and down into the lungs, with the assistance of the diaphragm. The diaphragm tugs down on the lungs to better improve their efficiency in driving oxygen molecules along a spaghetti junction of 1,500 miles of airways (compressed inside 1,000 square feet of lung tissue) to the billions of cells of the brain and muscles.

Atmospheric air, with which the sphere is filled, consists of 78 per cent nitrogen, 21 per cent oxygen, 0.93 per cent argon, 0.04 per cent carbon dioxide, and trace amounts of gases such as neon, helium, methane, krypton and hydrogen. For so small a component, CO_2 is crucial to both the life and death of a submariner. CO_2 is the gas that triggers the desire to breathe, that forces humans to inhale. This is counter-intuitive. We would think that falling oxygen levels are the primary respiratory stimulant, but anyone capable of holding their breath for three minutes will still maintain a normal level of blood oxygen. It will, however, be the rising levels of CO_2 that trigger the strong urge to breathe. Rapid exhalation is used to flush CO_2 out of the body and allow free divers to overcome the pressing desire to breathe, enabling them to go longer and deeper, but it can also lead to sudden unconsciousness and drowning if not carefully monitored. (The longest time a human has held their breath is currently 24 minutes and 3 seconds, by the Spanish diver Aleix Segura, which he achieved lying face down and motionless in a swimming pool.)

Human lungs can hold six litres of air at a time. Over the course of 24 hours we expel 2.3 lbs of CO_2 or the equivalent of 0.05 grams with each breath. Flatulence, incidentally, consists of 50 per cent CO_2, 40 per cent hydrogen; for one-third of people the final 10 per cent consists of methane. The odour is produced by hydrogen sulphide, present in trace amounts of 1 to 3 parts per million.

Yet it's CO_2 that's the silent assassin. Colourless, odourless and non-flammable, it will slowly build up in a confined space, accumulating more densely closer to the floor as it's one and a half times heavier than air. Its debilitating effects start gradually but rise rapidly the greater the concentration. Stepping into a room with CO_2 at 30 per cent would lead to almost instantaneous unconsciousness, with respiratory arrest within 60 seconds. Yet within *Pisces III*, it's more a case of boiling a frog: a slow and potentially deadly accumulation unless it can be kept at bay.

As CO_2 levels rise, the early symptoms are dizziness, drowsiness, a general fatigue, a shallowness of breath and flushing of the skin. Headaches are triggered by the development of hypercapnia as the CO_2 begins to change the pH level of the blood, making it too acidic. If this happens slowly, the kidneys begin to work harder, releasing and reabsorbing bicarbonate, a natural form of CO_2, which helps to keep the body's pH level balanced. Hypercapnia also leads to an increase in brain tissue and cerebral vasodilation, which triggers a debilitating intercranial pressure and headaches of remorseless ferocity (CO_2 causes blood vessels in the brain to widen).

When CO_2 rises above 5 per cent, there will be muscle twitches and involuntary hand flaps, depression, panic attacks and increased paranoia, and if it rises above 10 per cent there will be violent convulsions, followed by coma and death.

This precise, grimly unfolding sequence had recently occurred across the Atlantic, when a small submersible, *Johnson Sea Link*, was on a routine dive off the coast of Key West on 17 June 1973. The vessel became trapped in the wreckage of the American destroyer USS *Fred T. Berry*, which had been sunk to create an artificial reef and future tourist attraction. The *Johnson Sea Link* was designed to have two separate compartments: one for the pilot and the observer, and a second rear compartment capable of being compressed to the pressure of the current ocean depth. Two divers within this compartment could then swim out to carry out repairs and commercial maintenance. On the day of the accident the two men in the rear compartment were Edwin Clayton Link, a 31-year-old professional diver, and Albert Dennison Stover, 51, a submersible pilot. Neither man was diving that day and had dressed in shorts and T-shirts, despite a warning prior to the dive that it would be 'cold down there'. In the forward compartment were Archibald Menzies, 30, the pilot and Robert Meek, 27, an ichthyologist, who was in the observer's seat.

The plan had been to descend to the *Fred T. Berry* and use the mechanical arm to remove a fish trap, but when this proved too difficult the sub prepared to ascend. She then became caught in a steel cable attached to the sunken ship's wreckage. Rendered immobile at a depth of 360 feet, the pilot of *Johnson Sea Link* alerted the surface vessel *Sea Diver*, which notified the coastguard, and the naval ship USS *Tringa* was dispatched to the scene. The four occupants of the trapped sub believed they had sufficient oxygen supplies to wait until rescue, but the problem was the rise in CO_2 levels and a fatal variation in the temperature inside each of the two spheres. The pilot compartment was made of an acrylic plastic, which retained heat more efficiently than the aluminium hull of the rear divers' sphere. As the temperature in this sphere dropped to mirror that of the surrounding ocean, the baralyme within the scrubber, designed to absorb the CO_2, was rendered inefficient and CO_2 levels began to rise.

The *Tringa* arrived on site after seven hours but initial rescue attempts failed. A descent to 360 feet by two divers in hard hats was called off when they were unable to get close enough, blocked by the rusted hull of the *Berry*. Twice a lock-out dive was considered, but on the first occasion Link was reluctant to attempt to free the sub manually and by the second both divers were too cold to perform the task. The rising CO_2 levels rendered the atmosphere increasingly toxic to breathe, so Link and Stover were forced to wear diving helmets and breathe a mixture of helium and oxygen, which further reduced their body temperature. At 1.12 am, almost 15 hours after the submarine became trapped, Menzies and Meek reported from the other sphere that Link and Stover were suffering violent convulsions. After this, all attempts to communicate with the two men failed.

It was not until the following afternoon that a commercial salvage vessel – equipped with a remote-controlled underwater TV camera to which a grappling hook was temporarily attached – was able to reach the submarine and attach the hook to the sub's propeller. The submarine was then cautiously hoisted to the surface. Menzies and Meek were rescued alive, while the bodies of Link and Stover were recovered, the cause of death recorded as 'Respiratory Acidosis due to Carbon Dioxide Poisoning'.

If Mallinson and Chapman want to remain calm and limit oxygen use, they should aim to reduce their breaths to six per minute. The normal rate is between 12 and 16 breaths per minute. Breathing at such a slow rate widens blood vessels, reduces the heart rate and stimulates the vagus nerve, which controls a response known as 'rest and digest', which is the opposite of the 'fight or flight' mode of efficient panic. If they're capable of reducing their breaths down to just three per minute, they will find themselves dropping into what those who meditate recognise as a deeper state of consciousness in which theta brainwaves increase. Yet the fact is they can't get that low.

They're also mainly breathing through their nose, although mouth-breathing, frowned upon by many, would be more efficient in this particular situation as it utilises less oxygen. But it's a trade-off; breathing through the nose increases brain function, but how much do they need to operate a scrubber? Breathing through the nose is correct, as it's how the human body is designed, with the nasal hair and mucus purifying the air by straining out dust particles and germs. The nasal cavity also regulates the temperature of the air, warming it up or cooling it down as required by the environment. The reason nose breathing increases brain function is that it more readily stimulates neurons in the olfactory bulb, which then send signals to the hippocampus, the area of the brain responsible for memory and learning. Yet breathing through the nose also adds 50 per cent more air resistance compared with breathing through the mouth, giving the heart and lungs extra work, and so drawing in 20 per cent more oxygen.

In the early hours of Thursday morning Mallinson awakes from a light sleep and is immediately gripped by a headache of unusual intensity. He's also feeling groggy and far from his sharp and analytical self. In the dark of the sphere, he can feel the warmth of Chapman beside him, as well as his partner's heavy breathing. Chapman is still asleep and so Mallinson remains still, willing the pain in his head to pass. Later, when Chapman awakes, he too is suffering from a headache but one not as severe as Mallinson's. Both men know that headaches are an indicative symptom of rising levels of CO_2, but when Chapman checks the Ringrose CO_2 indicator, the reading is not unusually high. Since changing the canister the scrubber has been working effectively and they have been punctilious – more or less – about its operation.

So if it's not CO_2, what is causing the headaches? It's indicative of their addled state that it takes both men a considerable amount of time to recognise a change in their sleeping arrangements. If they lie down, their heads are now considerably lower than their feet. While asleep their blood is pooling in their head, triggering their headaches. At some point during the night the benches have either slipped or *Pisces III* has sunk deeper into the mud; either way, the situation needs to be fixed. Chapman nods back off to sleep as Mallinson sits up and starts to figure out what to do. As there's not much to hand that can be used to raise their heads by a dozen inches or so, it becomes apparent that they're going to have to leave the benches as they are, and instead both will have to move from one end to the other, exchanging head for toe.

The small dome lights are switched on, briefly dazzling Mallinson. When his eyes adjust to the light, he begins moving the equipment positioned by their heads to the other end of the benches. Chapman finds it's not possible to move the oxygen bottle, which was so conveniently by his head, but accepts this as a fair trade as he is now closer to the scrubber switch. The CO_2 contents gauge is re-secured to the wall near their heads, and each man moves his own handheld timer. The reorganising takes longer than they expect, over 30 minutes of movement, laboured breathing and much discussion. When they check the CO_2 indicator there has been a sharp rise. Chapman switches on the scrubber and runs it longer than normal to clean out the CO_2, then bleeds more oxygen back into the sphere. They agree that, from now on, they won't move or attempt to switch anything else around in the interests of comfort, only for reasons of safety.

As Chapman is now standing, he peers out through the observation port and notices that it's no longer a black mirror. Instead there's a very slight 'lightening' in the water: a wash of grey, which he records in his notebook. On the surface, dawn is breaking.

Dawn heralds more than a return of the sun. As the sky lightens from grey into a burnished gold, Ralph Henderson is on the bridge of the *Sir Tristram* looking towards the horizon and the imminent arrival of a second Royal Navy vessel. The HMS *Hecate* is a 2,898-ton long-distance survey ship, designed for deep ocean work and capable of 14 knots. She's carrying a crew of 14 officers, 104 ratings and a hold filled with exceptionally strong rope. At 5.45 am the grey hulk of *Hecate* is in position, and preparing for Henderson and Mayo to transfer across from *Sir Tristram*.

Before departing *Sir Tristram*, Henderson calls down to *Pisces III* to give a brief, optimistic update on the rescue plans. *Voyager* is due to arrive in Cork harbour anytime soon. *Pisces II* and *Pisces V* are already dockside, and the Americans are expected to be touching down later today. Henderson knows today will be another long wait for the pair. He also knows that the latest long-range weather forecast is not encouraging. The winds are picking up, a gale is now blowing and the weather from the west is deteriorating rapidly. He passes none of this information on to the men, but ends with an explanation that there will be another communication delay while they again transfer ships.

Henderson and Mayo are on *Hecate* by 8.39, where they have roughly the same communication set-up as before. Henderson has a brief communication check with *Pisces III* at 9.10 and afterwards he updates the base at Barrow, but explains that he hasn't yet got the latest level check on their oxygen supply and batteries, which he assures them he'll collect on the next call. Barrow gives Henderson the latest situation report on *Voyager*'s progress in Cork.

The deteriorating weather is now starting to affect communications with *Pisces III*. The heavier the waves slapping against the side of the ship, the greater the disturbance to the underwater telephone transducer, which is both microphone and receiver. The size of a large coffee mug, the transducer is lowered over the side of *Hecate* and hangs 15 feet beneath the surface, held in place by wire and a support rope.

In the bunk, Roger Mallinson is holding his head and breathing through the pain. The headache previously ascribed to the inverted sleeping position has returned and can no longer be blamed on the elevation of his feet above his head. The CO_2 in the atmosphere is rising, and clearly he is more susceptible to its effect than Chapman.

Chapman, meanwhile, is brooding in his bunk about the deteriorating communication with the surface. Often when they attempt to get through to the surface all they can hear is the chatter and squeaks of dolphins. This infuriates Chapman, but Mallinson finds it comforting, in spite of the problems the mammals present. Over the course of the morning, *Hecate* has failed to contact them on the hour and half-hour as agreed, and every missed call heightens Chapman's anxiety and sense of isolation. What if all communication from the surface ceased? This is the thought that preoccupies Chapman, though rather than simply worry, he prefers to worry and prepare. In the dark he pats around until he finds his small notebook and formulates a rudimentary communication system, writing: 'SIGNALS – ONE GRENADE EVERY HOUR. 3 WHEN RECOVERY OPS START.' He knows *Hecate* will have small explosives on board which, detonated at a depth of 20 feet, will be picked up by *Pisces III*'s underwater telephone as well as every sea creature within a radius of 20 miles or more. The idea provides some brief comfort, though when he discusses it with the surface later, they veto the plan.

Communication improves after 11 am, and when Henderson breaks the news that *Voyager* has left Cork harbour and is now on her way back to the accident site, Mallinson and Chapman share a quarter of a cup of cold coffee as a celebratory toast. The pair then pass on information to the surface: 'Oxygen 1 at 1500; 1 at 1900 CO2 1 at 24 hrs. 1 spare canister.' The food supplies are now a quarter of a flask of coffee, milk and sugar, and one can of lemonade.

What they do not know is that the rescue plan's outer edge of Saturday morning is based on both men reducing their oxygen consumption to 0.25 litres/min/man, which is below what they are currently consuming. As best as Chapman can calculate, he and Mallinson are around the level of 0.3 litres/min/man. In the knowledge that nothing is likely to happen before midnight, they again agree to attempt to spend the next twelve hours in silence – or as close as can be achieved. As Chapman later wrote:

Every unneeded word spoken would mean so many fewer seconds for the rescue, while every wasted movement might cut off minutes. Even thoughts and worries could steal survival time.

In the afternoon Chapman begins to suffer from leg cramps, a dull ache that quickly magnifies into short-lived shooting pains that can only be exorcised by twisting around or attempting to straighten both legs. This means moving further into Mallinson's personal space, as the pair are lying tightly together to share body heat. But Mallinson doesn't mind. In fact it's a comfort to know that he's not the only one in discomfort, as his headaches are growing increasingly severe.

Chapman would later describe the scene in some detail:

The effects of a CO_2 level continuously higher than normal were beginning to show. It was painful to sit up every half-hour and search for the oxygen supply, and even to talk on the underwater telephone. The timers, when they went off, became particularly annoying, with a very sharp ringing noise. It was like lying in bed during the winter in a cold room, suffering from 'flu, with an alarm clock going off every half hour to remind you of the reality of staying alive. Our reality was a little difficult to comprehend and face up to, as the easiest thing was to do nothing – just turn over and sleep through the next timer, and then no more noise to worry about. If we did this it would be more difficult to wake up the next time, after yet another 30 minutes of CO_2 had taken its effect.

At 2 pm Mallinson and Chapman test the underwater telephone on the emergency battery, just in case this became necessary to use. The back-up worked as planned, reversing straight away to normal supply. Of the six hours between 2 and 8 pm, Chapman notes: 'Morale fairly good, continuous headaches but not disastrous. Did not feel as tired during the day as the previous night. Probably purely psychological.'

CHAPTER EIGHT

They had gathered since before dawn on the docks of Cork harbour, anxiously awaiting the first sighting of *Voyager*. The huddle of reporters, photographers and TV crews are keen to speak to the captain and his men for the latest news on the two men. All morning the quayside has begun to fill up with a fleet of articulated lorries and flat-bed trucks carrying heaving equipment, giant coils of rope, and the unmissable red and orange shapes of *Pisces II* and *Pisces V*, recently brought in from Cork airport. (*Pisces V* had arrived at the airport at 3.30 am, *Pisces II* at 4.12.)

Ever since the news of the sinking of *Pisces III* broke the previous morning it has dominated the news bulletins on both TV and radio, and been splashed across the front pages of the nation's newspapers. Fleet Street has dispatched its best reporters and photographers to Barrow, as well as Windermere and Broughton, the home town and village of Mallinson and Chapman. In Cork the day rate for hiring a fishing trawler has risen dramatically as rival papers compete to secure the fastest boat out to the accident site. Among the gathered press is Harry Dempster, a renowned photographer with the *Daily Express*, who has no intention of hiring a trawler. He intends to get a lot closer to the story.

A handsome man with thinning, slicked-back hair and a slight resemblance to Laurence Olivier, Dempster wears a signet ring on his pinkie finger and uses Nikon cameras. Over the years, he's shot them all: the Beatles, Elizabeth Taylor and Richard Burton, and a young Ali MacGraw in a couple of fashion shoots. Witty, but with a short temper, Dempster has a prodigious appetite for alcohol. When accidentally kicked unconscious by a horse while covering the racing, so the legend went, the picture desk asked what a horse was doing in a pub. He is, however, a superb photographer with a fine eye for an arresting image.

Roger and Pamela Mallinson live with their children on Keldwyth Drive not far from the banks of Windermere in an A-frame timber house that Roger built himself. It doesn't take long for the press to find the address and Pamela – training to be a teacher now that the children are a little older – tries to be polite to one of the first reporters to ring the doorbell.

'All I can say is that my husband is doing his best to get it up,' she says. 'He's not the sort of man to panic. I know him too well. He's exceptionally well experienced and he's a very, very sensible person. Our three children have been told what has happened but they are not panicking, they know their father too well.'

The children will panic later, though not about their father. Instead they're troubled by the behaviour of the press. The front gate is soon locked to reporters but they begin to climb over the fence, ringing the doorbell and going around the house, rapping on the window to attract attention. A neighbour, troubled by the reporters' behaviour, uses the Mallinson family dog, a large Alsatian called Amber, to run them off the family property.

Pamela arranges to stay with the children in the home of a family friend. It's to this large detached house in the centre of Windermere that Maurice Byham is dispatched by Vickers Oceanics. Byham, an experienced submarine pilot, is on holiday in London with his family when he first hears of the sinking on a radio news bulletin while standing outside a tube station. Rushing to a call box, he contacts the office to offer his assistance and is asked to come straight back. So he and his wife Millie pack their disappointed young son back into the family Ford and head north. Roger's twin brother Miles Mallinson, currently on holiday in Germany, is also contacted by Vickers and flown home to help support Pamela and the children.

But when the first reporter arrives on the doorstep of Doris Mallinson's house on Prince's Road in Windermere she seems happy, almost relieved, to talk about her son. 'It's been terrible,' explains Doris, who has recently retired from teaching. 'I've had one or two worries in my life but I've never had anything like this. It's the waiting and not being able to do anything that's so awful. The one comfort I have is to know that if there's anything possible to be done, he'll do it. He's a very practical soul. He was always very good with his hands, but never very good at exams, although he always got there in the end.'

In an attempt to bridge the emotional gulf that has long since grown between mother and son, one now compounded by disaster, distance and depth, Doris conjures not the man Roger has become but the boy he once was. 'He will never grow up to me. I still look at him as a kid in infants. I think: "Poor little chap, what a place to be in!" Then suddenly I realise that he's a man.'

Her son's fate has had her 'trembling inside' and 'going over and over it in my mind'. Yet, as she explains, she does not doubt his determination. 'He'll leave no stone unturned, nothing will be too hard for him. There isn't a thing that he can't do. He's practical in the extreme. I had a feeling something like this might happen one day. But Roger is a very grand boy and very determined. I am sure he will keep calm and do everything possible to get the submarine back to the surface.'

By comparison with the experiences of Pamela and Doris Mallinson, June Chapman's exposure to the press is more tightly controlled. Protected within the cocoon of Vickers Oceanics, she gives a brief telephone interview to the Press Association, who syndicate her comments around all the national press.

'I think he will be standing up to the pressure very well,' she says. 'They teach you all about that in the navy. He is incredibly patient, tolerant and kind, and he has a terrific sense of humour – though I don't think he'll be cracking any jokes at the moment. He will be worried – they both will – but they will be conserving their energy and air, and they know everything is being done to help them. Nothing like this has ever happened to him before. There must have been crises when he was in the navy, but he was always able to sort those out himself. Naturally I'm terribly worried, but it's terrific to see what's being done. There's such cooperation. The Americans are involved and it gives me great hope.'

Maurice Byham arrives at the house in Windermere. Roger's twin brother Miles, whom Byham notes is really quite identical, even down to the wispy beard, is attempting to comfort Pamela, although it must be rather peculiar for her to be sitting there talking to what looks like Roger about what's going on. Byham has been told by headquarters to keep Pamela away from the TV news and radio reports, and to explain, calmly and rationally, what's being done to rescue her husband. But it's proving hard. At one point he finds himself having to lay out the worst-case scenario.

Byham, sitting in a low-slung chair and facing Pamela and Miles on the living-room sofa, begins to explain the process of asphyxiation and carbon dioxide poisoning. He finds it hard to concentrate as the presence of Miles, almost an exact replica of the man whose fate he is sealing with every sentence, is deeply disconcerting for him too: 'So I had to sit there and explain how he would die.'

As an experienced submarine pilot, Byham attempts to picture himself in the same situation, and begins by explaining how the air system works, how they use egg timers to dictate when to operate the barometer every 30 minutes or so, how they will take out the CO_2, and how you can watch the pressure drop, then turn on the oxygen valve and watch it go back up again. After painting a picture of the system, Byham explains that, slowly but surely, if they run out of oxygen they will suffocate.

* * *

At Cork harbour Captain Len Edwards is taking some time out from coordinating the loading of *Voyager* to talk to reporters and pose for a brief photograph. The ship arrived in port at 8.15 am, and since then everyone has been hard at work getting fresh supplies and both the submarines on board. Edwards begins by explaining what he believes happened: 'I am not sure whether the hawser attached to the tow line fouled. It had been normal recovery, but the sub went under the ship and when it was 170 feet beneath the water the tow line broke and it sank to the bottom.'

Edwards says he's told Chapman and Mallinson to 'lie down and shut up', then explains: 'It's so important for these men to preserve oxygen at all costs. At the moment they have plenty, but there's no sense in squandering it or tempting fate. The best thing they can do until we bring them up is rest and sleep.' He projects confidence, predicting a swift resolution that, he believes, will see the men on the surface by this time the following day: 'The two men are special friends of mine and I will be glad to welcome them aboard for breakfast.'

Next, Messervy gives a short briefing on how he expects the rescue operation to unfold: that one of the *Pisces* craft, carrying the new 'hook', will sink down to *Pisces III* and attach it: 'Once the hook is on *Pisces III* it should be a relatively simple operation for it to be hauled up.' He makes it clear that they have everything in their favour: good weather, the best equipment and, he believes, the luxury of time: 'We've got plenty of time. At the moment there's no danger to the men. They might be uncomfortable but they have full life support. On the surface we can cope with the swell and winds of up to Force 6 and 7. Winds above this might make the surface work, launching and final recovery tricky.'

Tricky, but eminently manageable. Messervy is eager to present a positive front, so he says that he certainly doesn't see any additional difficulty being caused by the extreme depth – which he knows is untrue – and that the presence of CURV-III is just a means of 'covering every contingency'. He finishes by assuring the press: 'Once we get on with this type of operation it will only take us a very short time,' words to make fate's face curl into a wry smile.

When another reporter speaks to Bob Starr, one of the Hyco team, the Canadian is keen to sing the praises of their miniature sub: 'We're using the most sophisticated equipment in the world; we're leaving nothing to chance.'

The only American talking to the press is Bob Moss, the deputy head of the US Navy salvage department and a veteran of both the Palomares hydrogen bomb recovery and the rescue of the *Johnson Sea Link*. He has a warm, crinkled face and a shock of white hair, and is happy to downplay the US Navy's involvement. 'From what I understand we probably won't be needed,' he tells the gathered reporters. 'The mini-subs can do the job. We are the back-up.' Yet, like a proud father of a precocious child, he's also anxious to advertise CURV-III's considerable abilities. 'If we're needed, we can work at 10,000 feet down in any sort of weather. CURV transmits TV pictures as good as those you get in your front room. It has SONAR, takes 35mm films and runs on three 10-horsepower motors. Its SONAR will tell you where something is. You can go right up to it and then spot it with the cameras. It's so easy to handle. It's like a kiddy car.'

The Irish dockers will take just over two hours to load both *Pisces V* and *Pisces II* onto *Voyager*, which is ready to sail at 10.35.

Al Trice is running late. He's not going to make it down to Cork harbour in time, and Captain Edwards insists that *Voyager* isn't going to wait for anyone. Trice arrives at Cork airport at 10.20 am, where he is met by a member of the Irish Air Force, who escorts him across the runway and onto a rescue helicopter. Trice isn't sure if *Voyager* has a helideck, and realises it doesn't when the crew explain that they won't be touching down but he will: lowered 100 feet on a harness. The notes in the daily log back in Barrow record that at 12.30, 'A. Trice dropped on *Voyager* (gently) by helicopter.'

Once on board, Trice is warmly welcomed by Messervy, who is delighted to see him and to have a reliable second-in-command by his side. Trice, however, wastes no time drilling into the current rescue plans. First he wants to know the current temperature on the seabed. He's told it's 54°F (12°C), which is a relief – reasonably mild and with little chance that both men will be suffering hypothermia, which can happen if the water temperature is close to 40°F (4°C). Then Trice asks about the specific current location of *Pisces III* and is 'dumbfounded' to discover that the submersible is without an active sonar pinger, and that the only pinger on the seabed is next to the cable, designed to guide the next working sub to the current work site. Hyco's newest subs such as *Pisces V* are next generation and fitted with a sonar transponder – a two-foot-long cylindrical tube six inches in diameter that sits on the coning tower and responds to a signal sent from the surface vessel. Vickers have been reliant on the use of surface buoys connected by rope to the submersible on the seabed, but, as Messervy explains, this was disconnected on the surface prior to the sinking. There is the emergency buoy that should have been released, but it has either malfunctioned or broken off during the descent. Although *Voyager* has sonar, she really has to be positioned right over the sub and to hit her directly to give an accurate location. *Voyager* isn't equipped with any dynamic positioning equipment, and because she's going to be pitching and rolling with the weather she will probably keep missing her target. It becomes clear to Trice during the conversation that no one knows exactly where *Pisces III* has landed.

Trice is not the only recent arrival on board *Voyager*. Harry Dempster, the *Daily Express* photographer, is now making himself at home. To this day it's unclear how he managed to get onto the ship. There were rumours that he sneaked on board at Cork harbour and managed to hide in one of the lifeboats until the ship set sail, at which point there was nothing the captain and crew could do, as turning back to unload a stowaway risked further, potentially fatal, delay. A Fleet Street veteran, Dempster knows it's usually easier to gain forgiveness than permission. Another possibility is that Dempster was 'invited' on board on a 'pool' basis and that all his photographs would be distributed by the Press Association to every media organisation that wished to use them, rather than for the exclusive use of the *Daily Express*. What is known is that at 6.15 pm the head office at Barrow succeed in getting a call out to *Voyager* to confirm Dempster's presence and to make sure that he's aware that any pictures will be syndicated by the Press Association. Back in Cork, Doug Huntington is tasked with ensuring that, upon Dempster's return, the photographer is not allowed to escape with any additional 'exclusives'.

The message back from *Voyager* as she drives hard through the grey waters of the Atlantic is that *Pisces II* is expected to be in the water by 1 o'clock early on Friday morning. The weather has picked up from the west and is blowing into their faces, whipping up waves and making progress slower than they had hoped.

The two American C-141 Starlifters touch down at Cork airport shortly before 5 pm on Thursday afternoon, and Larry Brady, for one, is as unimpressed by the reception committee as he is by the Irish weather, the balmy warmth of southern California now replaced by a blanket of cold, wet fog.

'Now I won't say that I expected a brass band or anything when we arrived,' he recalls, 'but the scruffy bunch of men waiting for us was a far cry from what I expected. The military airlift command personnel were dressed in civvies, especially the Air Force master sergeant, who was wearing a turtle neck, jeans and a Greek fisherman's hat.'

Sartorially, the Irish air crew may be a disappointment to Brady, but the heavy lifting gear is in place and the Irish team are hard workers. There's also a friendly face and a fellow countryman as Earl 'Curly' Lawrence, the premier salvage master for the US Navy, has flown in from Washington. Brady asks Curly where the ship is, and it's explained to him that the Canadian icebreaker the *John Cabot* is tied up ten miles downriver. The tide is out and it's going to keep her there for at least a few more hours.

Brady, Lawrence and Bob Watts, CURV III's programme manager, borrow a Volkswagen Beetle, and despite never having driven 'on the wrong side of the road' as they see it, set off – at speed – through the winding country roads so that Brady and Watts can inspect the *John Cabot* and see what the underwater recovery vessel will be operating off.

'Once we arrived, and I saw how far the 01 deck [the first deck above the main deck] of the *Cabot* was above the surface,' Brady says, 'my eyes rolled back. It was really way up in the sky. It sure as hell wasn't the *YFNX*, a special-purpose barge that is nice and close to the water. Two large cranes were on the deck and the vehicle would have to be lowered 35 feet before it hit the water. It was an accident waiting to happen.'

Curly and Brady look at each other. The *Cabot*'s captain wants them to make up their mind, but the fact is they don't have a choice. This is what they have to work with, but Brady wants to at least pretend that it's his call.

'I can live with this,' he says.

To make the tide, the *John Cabot* has to set sail now, so all the equipment is going to have to be sent out to the ship by barge. Brady is about to head back to the car when Curly stops him. 'No time,' he says bluntly. 'We'll work on board and lay out the equipment.'

CURV-III and all the equipment are ferried out to the *John Cabot* on pig barges. 'I had to smile as I watched Denny Holstein standing there in the mist, knee deep in pig shit,' says Brady. 'Well, maybe it was really ankle deep.' The barge's most recent passengers were clearly nervous passengers, to

CURV-III preparing to launch.

judge by the thick layer of steaming manure onto which their million dollar machine is about to be rested. The Irish crew grudgingly unfurl a long green hose and begin to spray down the deck, the fresh water sluicing the dung over the sides and into the bay.

Once all the equipment is on board the *John Cabot*, the team weld the control and supply vans down onto the deck to keep them rigidly in place during the sea voyage. The *John Cabot* has 15 miles of Nydac braided rope in spools of 1-inch, 2-inch and 3-inch diameter. They receive word that the *John Cabot* now won't set sail until 6.45 on Friday morning. This puts them roughly 18 hours behind *Voyager* and the British and Canadian teams, and the general feeling is that it will probably all be over by the time they arrive.

It's late afternoon in the boardroom of Vickers, and Sir Leonard Redshaw and Greg Mott are sitting behind two microphones resting on a long polished wooden desk as together they host a press conference. Time remains the critical concern, and Redshaw and Mott insist that the rescue operation is still operating to a hard deadline of 10 am on Saturday, when the oxygen supply in *Pisces III* is estimated to expire. Yet, as they make clear to the assembled journalists, they believe Chapman and Mallinson will by then have long since been rescued. In fact, their hope is that they will be on the surface, enjoying breakfast within 12 hours from now, a full day and night before their oxygen is expected to run out. Together the image both men project is one of smooth and assured confidence. The rescue will begin in the early hours of Friday morning and should only take a few hours, after which, 'Chapman and Mallinson should be able to have a hearty breakfast on board the mother ship.'

'We must not overlook the fact that you are always at risk under these circumstances,' Redshaw explains, 'but we have done simulated rescues. We have taken all the precautions we can see to safeguard the position.' He goes on to say that the rescue will take place in the dark. This isn't a problem, as all submarines operating at this depth do so in the dark, assisted with lights. The problem is not the absence of light but the lack of time. 'The main factor we need,' he stresses, 'is speed.'

Then there's the weather. Redshaw makes it clear that this is the reason they decided to put both *Pisces* submersibles on *Voyager*: 'The indications are that it should be reasonable for the next 48 hours, but it was not too predictable in that area.' He says that the forecast is for winds of Force 5, backing to Force 4 westerly, with an 11- to 12-foot swell, adding, that, of course, 'We would like it smoother.' Of Mallinson and Chapman he says, 'Knowing the type of men we are dealing with, we told them the programme of the rescue operation. We thought there was no point in them wondering when someone is going to come. It's better for them to know and mentally relax until this time.'

In an attempt to reassure the press of the men's current conditions, Sir Leonard gilds the lily by insisting they have more than adequate food and drink, and even means of entertainment: the tape recorder is available for their amusement and both men, he explains, are usually allowed to bring down two of their favourite musical tape cassettes.

When the presence of the Americans is raised, the use of CURV-III is, once again, downplayed and presented as a back-up plan, only to be used in the event of something going wrong with the initial rescue attempts. The *Daily Telegraph* quotes a 'Vickers spokesman' as saying, 'At the moment the optimism among the Vickers Oceanics team is very high. We feel confident we can do the job ourselves with the equipment we have got.'

The narrative Vickers is rolling out is that *Voyager* will be on site by early evening. The first dive will take place before midnight and the whole operation wrapped up within two to four hours. So optimistic is Vickers' tone that when preparing their report, the *Daily Telegraph* does so on the assumption that readers will learn of the rescue on the radio, over their cornflakes on Friday morning, and even refers to the rescue in the past tense:

After their rescue early today, the men were expected to go aboard the Vickers *Voyager* to be taken to Cork. It's assumed, if medically fit they will be flown from Ireland to Vickers's own landing strip at Walney Island, Barrow-in-Furness, sometime today.

The Times is more cautious with its reporting in its Friday-morning edition:

There were hopes that by this morning two sister craft of the sunken *Pisces III* would have attached a hawser and hauled her to the surface. The time limit for the operation is 10 am tomorrow, when the submariners' oxygen supply runs out.

At 9 pm on Thursday evening David Mayo on *Hecate* tries to contact *Pisces III*. He records the conversation.

'*Pisces*, *Pisces*. This is *Hecate*. Do you copy? Over.'

Mayo listens to a wash of noise, the squeaks of dolphins mixed with the churn of engines and what sounds like a tin can being dragged along a long gravel road, but there is no reply.

'Long pause here with no answer,' he comments on the audio, 'so I'm switching the tape off.'

Ten minutes later he tries again. There's an almost immediate response, although the words are punching up through a soup of other sounds.

'Surface. *Pisces*. Do you read me? Over.'

'Loud and clear. Sorry about the delay. We had surface and positioning problems.'

'Say again. Over,' says Roger Mallinson, who has the handset.

'I say again. Sorry about the delay in contacting you. We had trouble with porpoises and positions. Over.'

'Roger,' says Mallinson, but the rest of his words are lost in the churn of static and the surrounding sea.

'Wait. I have a long message for you. Here goes. "Regret misunderstanding over *Voyager*'s ETA. Current one is 00:45." Over.'

'Roger.'

'"However do not get disheartened. Everything is really hotting up. We are constantly in communication with *Voyager* and *Aeolus* about arrangements in hand." Over.'

'Roger,' says Mallinson, who then launches into a description of how he wants *Pisces III* to be picked up. Only parts of this are clearly audible: 'I would like to suggest that we are picked up in the horizontal position ... I think the chains would flex this way ... and one to the main lift point hook.'

'Roger. We will think about this and come back. I have further messages. *Voyager* is full of bodies doing their thing. Des D'Arcy and Roy Browne are coming for you in *PII*. *PV* is crewed and ready also. Over.'

'Roger. Sounds very good.'

'Yes. It's all happening. I also have messages from home for you both. Here goes. For Roger Mallinson: "Keep your pecker up." I spell the next word. Lima. Alpha. Delta. Delta. Yellow. Lima. Alpha. Delta. "Laddy Lad. Love you. Mum, Jennifer [his sister] and Miles." Over. Next one for Roger Chapman: "All family send love. Hope it won't be long before we see you. Sorry I can't be with you for breakfast. Love you. June." Over'

Chapman then takes the handset: 'We're both fine.'

PART IV

FRIDAY

PART IV

CHAPTER NINE

At night the Atlantic Ocean appears as an endless wash of dark grey, dotted by the white crests of waves visible in the moonlight. Four ships form a circle around a white plastic buoy, illuminated by the spotlights of HMS *Hecate*, like wagons corralled around a camp fire in the vastness of the Great Plains.

On the fringes of this latter-day corral bob a posse of smaller fishing vessels, hired by brown envelopes of cash at the quayside of Cork harbour and transformed into the floating newsroom of the world's press. In the wheelhouse the captains have tuned their shortwave radios to the frequency used by *Voyager*, and if lucky they can hear the voices from the deep, the brief conversations with Roger Mallinson and Roger Chapman, sunk nearly a third of a mile beneath their bows.

It's well after midnight – closer in fact to 1 o'clock on Friday morning – when *Voyager* returns to the scene. The return journey of fifteen hours, slower than hoped because of strong westerly winds that meant forcing through heavy oncoming waves, has been well spent prepping both *Pisces II* and *Pisces V*, and sketching the outlines of a rescue plan. As an experienced military man, Messervy is aware of the maxim, 'No plan survives first contact with the enemy,' and in this context the enemy is both time and a hostile environment. It's a maxim of which he will become increasingly conscious over the next 36 hours.

Ralph Henderson is relieved by *Voyager*'s return and is adamant that he should pilot the rescue submarine. He is quick to organise a dash in a Gemini RIB back across the dark, moonlit waters that separate *Hecate* and *Voyager*, one which turns into quite the ride considering the worsening weather. David Mayo is left on board *Hecate* to co-ordinate their response of behalf of Vickers Oceanics.

Once Henderson is back on board, he makes his way to the bridge, where Messervy and Trice are in conference. Henderson is happy to see Bob Eastaugh, who has raced across from the North Sea. As operations manager and a veteran of Vickers Oceanics, Eastaugh is, to Henderson's mind, the true leader, regardless of Messervy's rank.

Henderson makes the case that as the person in operational command of Dive No: 325, he himself should be responsible for the safe return of 'his team' by piloting one of the two *Pisces* craft. While Messervy listens and gives Henderson his due, it's a request that is quickly denied – and for logical reasons. Henderson hasn't slept during the last 48 hours, and while no one in the rescue party has had a solid eight hours, there are fresher men on board and ones equally if not more experienced.

At the moment the pilot of choice is on the deck, working under the spotlights on the final checklist in preparation for the launch of *Pisces II*. Thin-faced, with a mop of light-brown hair and wearing thick black horn-rimmed glasses, Desmond D'Arcy, known to everyone as 'Des', is the chief pilot at Vickers Oceanics and a man of quiet confidence. A graduate of Birmingham University with a degree in electrical engineering and underwater acoustics, before joining Oceanics he had spent two years in Sri Lanka teaching socially handicapped children. D'Arcy had been working with Bob Eastaugh on the North Sea job, and knows both Mallinson and Chapman well. He will be joined on the dive by Roy Browne, another *Oceanics* veteran. D'Arcy is fixing the rescue hook to *Pisces II*'s manipulator arm with tape. The toggle is attached to almost 4,000 feet of polypropylene rope 2.5 inches in diameter, which is stored in a large steel bin. *Voyager* has sailed with a choice of ropes, rising from 1 inch in diameter, all capable of lifting the weight of *Pisces III*, as even the thinnest line has a breaking strength of 11 tons. The best calculation is that *Pisces III* is now carrying an excess weight of between 0.95 and 1.5 tons due to the flooded aft chamber.

Messervy has decided that the operation should go with 2.5-inch-diameter rope as it offers the greatest breaking strength and is also more buoyant. It's thought that this will allow it to 'stream' up towards the surface and so not snag on the propulsion motors or lifting hooks of either submarine. Messervy's hope is to get the sub up in one swift operation, and if D'Arcy can land the toggle just right they could conceivably have Mallinson and Chapman back on deck before dawn breaks.

When D'Arcy indicates the sub is ready to go, Messervy orders that the Gemini RIB be readied in preparation to tow the sub out to the buoy marking the spot of *Pisces III*'s last descent. When lowered into the water, the Gemini is immediately battered by the waves, as the diver pulls on the rip to ignite the outboard motor.

It fails to catch.

He tries again.

It fails to catch.

Again. Again. Again.

There's something clearly wrong with the motor, so after more than ten minutes the crew decide to switch to a second Gemini craft while the engine on the first is examined on deck by the mechanics. But the same fault occurs with the second craft. The motor repeatedly fails to start and, in time, they move on to a third engine, which also fails to start. For almost an hour, *Pisces II* is left suspended on the A frame awaiting word that the dive can commence, with all attention focused on fixing three outboard motors that appear to have become contaminated with salt water.

The weather is growing increasingly foul. The wind is whipping up the waves and sea spray is blowing over the decks, where men in sodden boilersuits are either working or staring out into the darkness to where the buoy is being battered and bounced. Everyone knows that under normal circumstances there would be no launch tonight and any plans would be postponed as too dangerous. But everyone also knows that's not an option, even though the most vulnerable link in the chain during weather like this is the craft on which so much depends – the Gemini. Open to the elements, the small inflatable rubber boat can easily be flipped by a sudden encounter with the wrong combination of heavy waves and strong gusts of wind.

On the bridge, Captain Len Edwards is anxious to see the Gemini launched so he can manoeuvre *Voyager* away from the buoys, as holding the ship in position is far from easy given the weather. He waits as long as he can, then announces that they are pulling back until the engines are fixed. The problem is finally resolved when a working engine is sent over by another Gemini from HMS *Hecate*.

At 2.14 am *Pisces II* is lowered into the water – finally towed out by a Gemini. She is then unhooked, and by 2.30 has opened her vents to the sea and begins to sink. In the sphere, Des D'Arcy activates the lights and gazes out of the porthole as black water washes over the steel frame and fills his view with crud. He can see the rescue hook is firmly secured to the manipulator arm and the 2.5-inch polypropylene rope drifting back and to the left of the sub. It looks like a long grey tentacle, snaking back up to the surface where the crew watch as it uncoils and spools out of its own accord from the giant steel bin on the edge of *Voyager*'s sea deck.

D'Arcy knows that *Pisces III* is at a depth of 1,575 feet and that the position where the sub was last deployed was 1,625 feet. The rescue team know from Chapman and Mallinson that the depth increases the further west of travel so they believe the sub to be lying east of the pinger, but perhaps still close to the transatlantic cable.

In the sphere of *Pisces III* Mallinson is standing up, straining to see out of the sub's new porthole 'skylights'. The first indication of *Pisces II*'s arrival on the bottom and in the area nearest to them will be the faintest lightening of the water's inky black wash. He's anxious to see it, but Chapman can sense that the mix of excitement and anxiety is like an electrical current in Mallinson, burning up more oxygen, so he stands up, gently puts a hand on Mallinson's shoulder, who, sensing his concern, returns to lie down on the bunk. Chapman lies beside him, clutching the handset of the underwater telephone as they listen to D'Arcy and *Pisces II*'s progress.

The sharp din of the pinger is all they can hear for now and it has become a taunt. If only they knew their location in relation to the pinger, their discovery would be assured. *Pisces II* would be able to follow the din of the pinger, then head in the opposite direction of *Pisces III*'s bearing. Yet the stubborn fact remains that this is not an option. It may sound as if the pinger is almost outside the window, but in reality it could be a mile away in almost any direction.

'*Voyager, Voyager*, this is *Pisces II*. Communication check, over.'

'100 feet.'

'150 feet.'

'200 feet.'

'300 feet.'

'400 feet.'

'500 feet.'

The voice of Des D'Arcy is, for now, coming in slow and strong to the radio room on *Voyager*, where Messervy and Al Trice are leaning on the arched metal doorway as Ralph Henderson mans the controls. At around the 1,000-foot mark, about 20 minutes into the dive, the static and crackle drown D'Arcy's voice, and Henderson is left speaking into an empty mike. Yet in an acoustic twist, while *Pisces II* is rendered briefly inaudible, Chapman and Mallinson can hear every word from the sister sub and relay their communications to the surface.

'*Voyager*, this is *Pisces III*. *Pisces II* reports her depth at 1,000 feet. Over.'

In *Pisces II*, D'Arcy is delighted to hear Chapman's voice and joins in the conversation, but what the trapped men can hear is broken, something closer to: 'good ... hear ... voice ... will ... you ... shortly'. They also recognise that this is not sound coming from a submarine that is nearby, and the view out of the porthole gives no indication of a searchlight, however faint or distant. The rising hope in both men's heart is soon to crest and break.

In the cockpit of *Pisces II*, D'Arcy is lying flat on the wooden bench and looking out of the porthole. He can see the manipulator arm begin to flinch and strain. The 'T' bar rescue toggle is still taped in place, but the long grey 2.5-inch polypropylene rope is tugging hard at the mechanical arm. At a depth of 1,250 feet, *Pisces II* is dragging down an umbilical cord that's now over 2,000 feet long, when the submarine's drift from *Voyager*, to which she's tethered, is factored in. The rope is also proving more buoyant than anyone on the surface considered, and the longer the length of submerged rope, the greater the tension required to drag its full length down. Instead of steadily sinking as it trails behind *Pisces II*, the rope's thickness and buoyancy mean sections are drifting up towards the surface, like the hump in a big dipper. The deeper they go, the greater the tension and shake on the manipulator arm. The rescue rope has also been tied to the metal frame on the top of the sub to take some of the strain of the surface pull, but this tie has now broken.

D'Arcy is an experienced pilot. Staring out through the toughened glass at the manipulator arm, he knows when a part is going to give. The light lashings binding the rescue rope to the arm begin to pop one at a time. A second or two later, the tension on the line buckles the manipulator arm. The rope and rescue toggle are ripped out of the arm and pull away into the blackness. The sub is still 400 feet from the bottom.

Aware that Chapman and Mallinson are listening, D'Arcy is restrained in his report, stating only that the lashings are breaking away at 1,250 feet.

On the surface, Messervy requires no further details to know the rescue, or this initial phase at least, is in trouble. Trice, though disappointed, shows little or no emotion. He knows every operation is a push/pull of triumphs and setbacks, and that it's about working with what you've got, not what you want. They may have wanted a sub with an operational manipulator arm and an effective rescue rope but they have neither. What they do have is a sub almost at the bottom and a target whose exact location remains unknown. Messervy could recall *Pisces II*, but makes a quick decision to push on.

D'Arcy and Browne are told to continue to the bottom and await further instructions.

Bathed only in the faint green glow from the luminescent face of the depth gauge, Chapman and Mallinson lie in their bunks and listen as the depth count from *Pisces II* goes deeper. Although they are unaware of what was lashed to the sub, they know that anything breaking away at 1,250 feet is not good. But they're still relieved that the sub is continuing her descent.

Communication between *Voyager* and *Pisces II* has improved, so Chapman remains silent. He finds this easy, as his lethargy is rising, along with the CO_2. They haven't run the scrubber for 45 minutes as the noise makes it hard to hear communication from the surface and during that time both men have moved and spoken more than they have done in the preceding few hours. A quick look at the Ringrose indicator shows it has come at a cost.

They switch on the scrubber and wait.

The reddish form of *Pisces II* settles down on the seabed at 3.40 am, disturbing the sandy bottom and sending up a curtain of sand grains that turns the waters milky. D'Arcy and Browne are without the long umbilical rescue line, which is now being reeled back onto the storm-lashed deck of *Voyager*. The job is no longer to rescue *Pisces III* but simply to find her. If all the water in which *Pisces II* sits were removed, they would be looking out at a rough terrain of small hills and valleys, long flat stretches and sudden deep rivulets, all enfolded in permanent night. The pair know that somewhere out there *Pisces III* – the size of a small minibus – is standing upright, wedged on her tail. But where?

On each submersible is a sonar dome, a small, curved glass dome that hangs down just above the central porthole at the bow. As *Pisces III* is standing up with her stern wedged into the seabed, the sonar dome is sitting horizontal. A separate pinger receiver, shaped like a long thin torch, is suspended beside the submarine's cluster of high-powered lights.

In *Pisces II* D'Arcy and Browne are using sonar – sound navigational ranging. The system operates by sending out a pulse of sound that travels out as an ever-widening cone. Wide-beam scanning, using a span of 40° to 60°, is designed for quickly scanning a large area, while narrow-beam scanning, using a tighter arc of 10° to 20°, is designed to give a more precise picture. Sound travels through water faster than through air, at a speed of roughly one mile per second. If the arc of sound from *Pisces II* hits *Pisces III*'s transponder it will emit its own signal, one that will travel back to *Pisces II*. The time taken to receive the response will reveal the distance between the two submarines.

D'Arcy stares at a small green screen, seven inches by eight, on which is his 'sonic' field of vision. His hands are on the controls as he gently adjusts the motors and propellers so that the submarine rises a few feet above the surface, then slowly pirouettes around in a full 360° circle. All the while he is moving the sonar beam up and down in an attempt to make the coverage as comprehensive as possible.

To be in a submarine that is subject to a search by sonar is like being inside a clock. Chapman and Mallinson listen to the steady metronomic 'tick … tick … tick' of their sister sub's sonar. Yet both are experienced enough to know that the louder the tick, the closer the sub, and it's currently barely audible. The exact distance may be uncertain, but it's clear that *Pisces II* is a long way off.

D'Arcy is aware of exactly how far when the sonar picks up the faintest signal of what can only be *Pisces III*. The time taken to 'reply' is over a second, translated by the sonar device into a distance of 6,000 feet or around 1.1 miles – a huge distance when a difference of a few hundred feet can take a considerable time to pin down. D'Arcy and Browne are juggling the mixed emotions of relief that a positive result has occurred with the disappointment that they are clearly so far off course when they notice that water is beginning to drip down into the sphere. They have not been down long enough for condensation to build to the point where it drops from the ceiling, and a quick taste test reveals it to be salt water.

After first having the rescue line ripped from her grasp, *Pisces II* has now sprung a leak. The rubber seal around one of the penetrators is failing and the Atlantic Ocean is being forced inside the submarine at a depth of around 1,600 feet. The leak begins as a few drops, then develops into a steady 'weep' of salt water. D'Arcy, looking away from the sonar up to the source of the leak, knows he cannot remain in position. The search has to be abandoned and they begin an emergency ascent. He relays the stark facts to *Voyager*.

On the surface, Messervy tells Trice to check on the launch preparations for the Canadian team and *Pisces V*, as it's clear they're going to need to get the craft in the water rather sooner than planned.

D'Arcy begins pumping oil into the submersible's ballast bags to move into positive buoyancy. The sub lifts up off the seabed, the lights catching the movement of a passing shoal of fish, which quickly darts off into the darkness. A numeric count of the ascent begins over the underwater telephone, like a bellhop shouting out every floor of an elevator's rise: 1,500 feet, 1,400 feet, 1,300 feet, 1,200 …

Mallinson and Chapman are in a drowsy state as they listen to their rescuers' retreat. The scrubber has not been used for over an hour. The rise of CO_2 has made them long for sleep, though it cannot fully dull the stabbing pain of disappointment. The pair can make out the voice of Des D'Arcy through the static as he calls out the final few hundred feet before his voice fades, lost among the surface wash of the waves. Then all they hear is the pinger, always the pinger, dispassionately recording the passage of time in one and a half second delays.

The humidity in the sphere is at 99 per cent, according to the gauge. The lithium hydroxide pellets in the scrubber are once again failing, as both men are breathing out their CO_2, which the pellets inside are no longer able to absorb. This is the end of the second canister. It has been in use for 29 hours, the equivalent of 58 'man hours'.

In the background they suddenly hear another noise, not the automatic, mechanical din of a man-made location device but what sounds like joyful squeaks and inquisitive clicks. Mallinson is the first to recognise the sound, and he smiles. It's dolphins. Rationally he knows they can offer no assistance, but he cannot help but take comfort in their continued presence. Why would the pod not have swam on? He imagines what it would be like if one of them swam down this deep, its face appearing at one of the portholes. 'Wouldn't that be wonderful,' he thinks, briefly lost in a reverie of dolphins dancing to the music of Mahler.

Mallinson also notices a strange effect when he and Chapman attempt to contact the surface. It seemed as if the sound waves 'upset the phosphorescence in the sea'. As he recalls:

We were looking up into blackness. You could see all the phosphorescence going past with the tide. The stuff that was close was going fast and the stuff further away was going more slowly. It was like looking up at the stars at night or passing through the stars. It was as if you were up in the Heavens. There were just flashes of light, going past like stars and that only happened when you were transmitting.

Pisces II surfaces just before dawn into the darkness of a heaving sea, with the waves slapping hard against her hull. The surrounding ships pour their spotlights onto the sub and for a few seconds Bob Hanley, sitting in a black wetsuit at the front of the Gemini, is dazzled. He has exchanged his warm bath at home on Walney Island for the icy chill of the Atlantic to assist in the sub's recovery. The conditions are as difficult as he has ever experienced, and it takes longer than usual to connect the tow line, but finally *Pisces II* is hooked up and towed back to *Voyager*, where the steel A frame is lowered down to lift the sub back onto the deck.

The hatch opens at 4.18 am. D'Arcy and Browne climb out, joining the technicians who are already examining the damaged manipulator arm. The force of the rope has bent the arm, while the hydraulic elbow ram plunger has been twisted. The whole unit needs to be stripped down and replaced before the sub is operational again. The two men leave the deck and head to one of the cabins off the bridge, which Messervy and Trice have turned into their joint command post. They are accompanied by Bob Eastaugh, operations manager. No one had considered the possibility of the rope ripping free, and the discussion focuses on when the rope's buoyancy began to cause trouble. D'Arcy says it was about 1,000 feet. The buoyancy of the combined length of 2.5-inch rope at that depth was too great for the straps. The team know that next time they will have to go down with a thinner rope.

The next topic of discussion is *Pisces III*'s most probable location. The buoy currently marking where the sub sank provides no indication as to where she eventually landed. D'Arcy talks through where he got the reading, but given that it was at a distance of 6,000 feet, everyone around the table knows that they are still pointing at a haystack and saying the needle is definitely in there.

Morale on board *Voyager* has been dented, but the team is conscious of how much greater the despondency will be below the waves. To ensure the two men do not feel forgotten, the communication check is increased in frequency from every 30 minutes to every 15 minutes, with Chapman or Mallinson required to click twice on the transmitter in response, as this limits speech and therefore oxygen depletion. The increased checks should help them to fight off sleep, although this is a losing battle. After responding with two clicks at 4.30 am, Chapman and Mallinson begin to drift off.

On the deck of *Voyager* the baton is passed from Britain to Canada. As the technical team from Vickers Oceanics start what they queasily know is shaping up to be a lengthy repair job on the manipulator arm of *Pisces II*, a dozen or so feet away the team from Hyco are running through the final system checks on *Pisces V*. The men work side by side as puddles of rain slosh across the deck and *Voyager* pitches and rolls with the waves. Messervy's plan is that each national team is run independently, as they best know their own equipment, systems and men, but that they're all under his overall command.

Trice puts Al Witcombe in charge of communications between *Voyager* and *Pisces V*. Witcombe bases himself in the control room, which is dominated by the large chart table on which weary crew have been known to grab a quick nap. The rough weather means he has to wind his legs under the table and hold on to the telephone to stop himself from tumbling over. When it gets really rough, the chairs are tossed to the floor.

Pisces V will be manned by Mike Macdonald and Jim McBeth, the two men with the greatest relevant experience. It was Macdonald, after all, who manned the submersible that rescued Messervy. Their plan is to borrow from the same playbook. They are going to use the same type of rescue line, but thinner and less buoyant than the 2.5-inch line used by *Pisces II*. Their rope is six-part braided polypropylene with a diameter of 1.5 inches and a commensurate breaking strength. They're not going to use the rescue toggle designed by the boys back in Barrow to fit into the flooded aft chamber; instead they will use an open crane hook with a spring-loaded gate and attempt to fit it onto *Pisces III*'s own steel lifting eye, the curved hole that sits just behind the main hatch and sail.

This is the way they did it down in the dark waters off Vancouver, and there's no reason why it shouldn't work in the depths of the Atlantic. During *Voyager*'s return journey from Cork harbour to the accident site, Trice and the Canadian team used the time to customise the snap hook. A steel lifting eye, a duplicate of the one fixed to *Pisces III*, was found on board, and they checked the snap hook to ensure a snug fit. It did fit, but only just, so to provide a smoother passage the team ground down the hook's bill to make sure it would slip more easily past the safety snap.

The snap hook is fixed to the rope, then fastened onto the sub's manipulator arm, and *Pisces V* is readied for launch.

In the wheelhouse Captain Edwards cautiously steers *Voyager* back towards the buoys, but he is uneasy. The ship's own sonar has a rough reading for the location of *Pisces III*, but surface sonar can be unreliable as the first three feet of water, the surface churn of waves, can be disruptive. Then there's the buoyancy system of *Pisces V*: a more recent model, the submarine's design has been tweaked and it takes longer to prepare to dive. On a calm, still sea this would prove no problem as the sub remains fixed in position, but in a stormy sea like today the risk increases that she will be buffeted and blown further off course. Uncertainty piles on top of uncertainty.

At 5.45 am *Pisces V* launches, and a few minutes later slips beneath the waves and out of sight of the Gemini and Al Trice, standing on the deck of *Voyager*, framed against a rose-coloured sky and breaking dawn.

* * *

The word that tugs Chapman from the comfort of sleep is 'metres'. He and Mallinson hear the first communication between *Pisces V* and *Voyager*, and it annoys them both.

'This is *Pisces V*, left surface OK, depth 50 metres.'

Why are they measuring in metres? What's wrong with feet, and for that matter, how many feet are there in a metre? When Britain joined the European Common Market on 1 January, just under nine months ago, the government insisted that beer would still be served in pints not litres and distance would be measured in miles not kilometres. Someone has clearly forgotten to tell the Canadians.

Questions such as how many feet are in a metre, posed on the surface and in safety, would be of no concern, but to those who are sleep-addled and tortured by headaches of increasing ferocity, the need to do a spot of mental arithmetic is an additional hardship. Chapman says it's roughly three feet to a metre, but Mallinson corrects him. He knows one metre is 3.28 feet, and if they don't keep track of all those extra 0.28s they could add up to the difference between remaining lost and being found. At 50 metres, *Pisces V* is not at 150 feet but 164 feet. From now on Chapman lets Mallinson handle the multiplication. It will give him something to focus on and stop him from getting overexcited.

On the descent, Macdonald's focus is fixed on the manipulator arm, the snap hook and the trailing rope to which it's attached. The previous dive failed because of the buoyancy of the rope, and while he knows their rope should cope, he still stares at it intently the deeper they sink, waiting to see if the whitish grey tentacle will move from dormant slumber and begin to kick.

The descent takes 30 minutes. The rope remains firm. *Pisces V* touches bottom at 6.15 am.

Macdonald activates the sonar and begins a wide scan. Within ten minutes of arrival, *Pisces V* has identified a target of comparable size to *Pisces III* and begins to slowly move across the seabed, hovering a few feet above the sandy floor, the bright lights illuminating a dozen or so feet (or four metres) of the underwater landscape.

A few hundred feet closer, and it becomes clear to Macdonald and McBeth that this is not their target. Chapman and Mallinson would sound a lot clearer by now, and have already responded to the lights of *Pisces V*. It's time to stop and begin again. The sonar begins to scan, focuses on a new target. They lift off the surface and head in a different direction. Again the target turns out not to be *Pisces III*. Once more it's time to stop and start over. Sonar. Scan. Wait. Watch. Identify target. Move off.

Chapman asks *Pisces V* for her current depth. The answer comes back – 460 metres. Mallinson translates this to 1,508 feet. Where *Pisces III* sits the depth is 1,575 feet, the equivalent of 480 metres. *Pisces V* is still in too shallow waters and needs to follow the contours of the seabed down onto deeper terrain. The deeper waters lies west, in which direction Macdonald now steers the sub.

Macdonald has an idea. As *Pisces V* slowly heads west, the sonar scanning as she goes, he asks *Pisces III* to start counting, slowly and strongly, from 1 all the way up to 50.

In *Pisces III* Chapman, who is already holding the handset mike, starts to count. The count will come at a cost in oxygen, but it's a bet worth laying.

1 ... 2 ... 3 ... 4 ... 5 ... 6 ... 7 ... 8 ... 9 ... 10

In *Pisces V* Macdonald has now stopped the sub from moving forward and is instead remaining stationary while slowly, very slowly turning the sub in a circle of 360° while listening over the headphones to Chapman's count.

11 ... 12 ... 13 ... 14 ... 15 ... 16 ... 17 ... 18 ... 19 ... 20

Macdonald can hear the numbers and Chapman's voice, and his hope is that just as he turns the sub another few degrees a specific number will suddenly rise in tone and intonation, punch through the litany of numbers and identify in which direction they lie.

21 ... 22 ... 23 ... 24 ... 25 ... 26 ... 27 ... 28 ... 29 ... 30

Macdonald listens hard, willing a number to rise in pitch.

31 ... 32 ... 33 ... 34 ... 35 ... 36 ... 37 ... 38 ... 39 ... 40.

In *Pisces III* Mallinson watches Chapman and briefly holds his breath, as if this will buy his partner better luck.

41 ... 42 ... 43 ... 44 ... 45 ... 46 ... 47 ... 48 ... 49 ... 50.

The count ends, but it's clear that Chapman's voice hasn't been able to smack through the muffled distortion of the sea. Macdonald asks him to try again and so the count begins afresh.

1 ... 2 ... 3 ...

By the time the second count ends as unsuccessfully as the first, Chapman's breathing is laboured and he feels light-headed. Mallinson takes over the handset. He's been feeling increasingly angry at Macdonald, who clearly isn't listening. Each time Mallinson has tried to brief them on how he thinks they should proceed, either Macdonald or McBeth has shouted him down and said there are too many voices on the line. Yes, thinks Mallinson, there are too many voices – but you need to listen to this one.

Mallinson has had plenty of time to think. He knows the only fixed point is the pinger dropped beside the cable upon completion of their shift. He also knows *Pisces III*'s ascent from the pinger took 45 minutes to reach the surface. As the submarine rose he had noted that all the sea flotsam and crud were running east with the ocean's tidal current, so when *Pisces III* surfaced she had to be roughly 45 minutes east of the pinger. Then when she sank, *Pisces III* would again have been carried east, so they had to be lying at the equivalent of the third point of a triangle. If the Canadians had any hope of finding them they had to first find the pinger, use it as their starting point and then push east.

Mallinson makes this point repeatedly to both *Voyager* and *Pisces V*. At 7.30 am Al Trice, in consultation with Messervy, tells *Pisces V* to head to the pinger.

At this point the *John Cabot* has only just departed with the American team and CURV-III on board. The men on the deck of the Canadian icebreaker, when not heads down working, look up at the green tapestry of County Cork as the landscape slips behind them at a rate of knots.

One hundred and fifty miles west and 1,500 feet down, *Pisces V* reaches the pinger at around 8 am, the submarine's lights picking up the device and the dark length of the transatlantic cable. The sub pirouettes around using the sonar in all directions, before concentrating on an eastward direction. As Macdonald and McBeth begin their search yet again, communications from *Voyager* become increasingly fragmented. At one point Macdonald reckons he's hearing about 1 per cent of what's being said. Macdonald begins to think, 'Jesus, we must be way off bat.'

CHAPTER TEN

On the bridge of *Voyager* Captain Len Edwards looks out through the panoramic windows, now smeared with rain. He doesn't like what he can see. A cluster of 50-foot and 75-foot fishing vessels, as many as half a dozen, are riding the high waves and pushing through the boundary set up by the positioning of *Voyager*, HMS *Hecate* and USS *Aeolus*, an American naval vessel, a veteran of the Second World War and now used for cable-laying.

Edwards knows each fishing boat has been hired by a different newspaper or broadcaster and that they are listening in to the VHF communications passing between *Voyager*, *Hecate* and *Aeolus*, as well as messages to and from *Pisces III* and the rescue subs. He also knows that their engines and propellers are increasing the audible churn, making communications between *Voyager* and *Pisces III* – already difficult due to the weather – even more problematic.

Over the last few hours Edwards has repeatedly ordered, over shortwave radio, that each fishing vessel move back, telling them that their behaviour is actively hindering a maritime rescue. There has been no response. So Edwards, in consultation with Messervy, speaks to the captain of HMS

Hecate, who agrees to dispatch the vessel's helicopter to deliver what they hope will be a final warning. A giant blackboard with a warning and order to back off chalked in white is attached by rope to the helicopter's undercarriage. After take-off, the helicopter arcs low over the sea and then swoops by each of the vessels in turn, hovering so the sign is clearly visible on the bridge. At last there's a response. The crew of the *Marina* come out on deck, both hands held aloft and the 'V' sign defiantly displayed. The Royal Navy is not going to tell an Irish fisherman what to do in international waters. The pilot cannot hear what they are shouting, but reports that it's most likely 'Fuck off'.

On board the *Marina* is Simon Dring, a tall, thin 29-year-old correspondent for the BBC, who is more used to stand-offs with the military of foreign governments than his own. By the age of 17 he had already hitchhiked across Europe, the Middle East and South East Asia, where he became the youngest correspondent with Reuters, based in Hanoi and covering the early years of the war in Vietnam. Most recently, Dring risked his life to stay in Dhaka in Bangladesh after being ordered to leave, in order to report on the Pakistan Army's massacre of civilians. It's when Dring's told to leave or back off that he's most likely to lean forward. He has a story to cover and that's what he's going to do, although it will not be easy. The heavy swells induce a bout of sea-sickness, and during a live two-way conversation with Terry Wogan, presenter of *The Breakfast Show* on BBC Radio 2, Dring accidentally vomits.

Edwards is furious. He understands the public interest in a rescue of this scale and complexity, and has tried to be

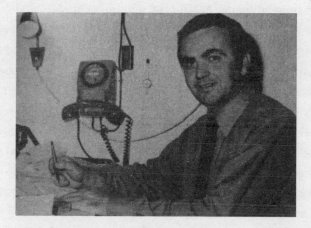

Captain Len Edwards, master of Vickers *Voyager*.

supportive when safe to do so. He spoke to the press on the harbourside at Cork, talked them through what they hoped to do and expected some cooperation in response. Instead they are no longer spectators, but – collectively – he believes, another barrier blocking the rescue of his men. A complaint about the press's behaviour is sent over to David Mayo on *Hecate*, who attempts to communicate it back to the Vickers' base in Barrow, in the hope that Sir Leonard Redshaw can exert some influence behind the scenes.

Edwards can live with factual errors. Today's newspapers are reporting that it's Roger Mallinson's birthday, and that his family have been permitted to speak to him, in case he does not surface alive. The source appears to have been the *Daily Express* photographer Harry Dempster, who misread a file on the ship detailing everyone's name, address and date of birth. Mallinson's date of birth is 31 March, not 31 August – a hastily scrawled '3' has been mistaken for an '8', and a false but emotive story is off around the world. Yet Edwards will not accept behaviour that he believes could have fatal consequences.

The buoy marking the spot is like the white steel hands of a clock fixed at ten minutes to two. To Edwards and others it's a constant reminder of time. On the bottom of *Voyager* is a scanning sonar designed to detect the submersible below. The position is then registered on a cathode ray oscilloscope in the operations control room, which gives the bearing and slant range of the submarine relative to the support ship. Edwards is increasingly convinced the ship's sonar is off and that he should trust his own instincts, maritime memory and experience to judge where *Pisces III* is lying.

At 8 am *Pisces V* moved to the point where the pinger sits next to the transatlantic cable. It's now almost 9.30 and still they haven't found *Pisces III*. Edwards tells Messervy once again that he wants to try a new spot, that they should bring *Pisces V* back to the surface and tow her over. Messervy finally agrees to give it a try, and at 9.47, four hours after *Pisces V* was first launched, she receives the order to return to the surface. Two different submarines have now dived, and both have returned without success.

Pisces V surfaces into high waves. A diver in one of the Gemini boats clambers on board but struggles to stay upright as the submarine pitches and rolls. The diver has several tasks: to connect a tow line and ensure that the snap hook gripped in the sub's manipulator arm remains in place, and making certain that the long polypropylene rope, trailing along on the surface, stays good and slack and does not get tangled, caught or cut by the Gemini's propeller blades.

Inside the sub, Macdonald reports to Trice that the submarine's gyro compass is acting up, with the needle wandering at a rate of 30'/hour, which makes direction from the surface difficult, if not impossible. Both men know that to uncover the source of the problem would require hauling *Pisces V* back on board *Voyager* and stripping the gyroscope down, which would take too much time. The gyroscope is a valuable aid but not as vital as sonar, which is still operational, and so Trice decides that they should push on. Once the straps are fitted, *Pisces V* is towed first by the Gemini back to *Voyager*, and then by the larger ship for roughly one mile. Edwards, Messervy and Trice are probably all in agreement that, if Mallinson's memory is correct about the direction of the tidal current on Wednesday morning, then the spot to which they are heading could be the third and final point on his triangle.

One mile distant and 1,575 feet down, Mallinson is no longer thinking of triangles. Curled in a ball, he is lying face down, shivering and shaking under the black plastic cover, and all he can think about is his headache, the worst he has ever endured. What feels like an iron band wrapped around the base of his skull and slowly tightening is the physical effects of rising carbon dioxide. Then there's the pain in his joints, spasms that sear then recede and then sear and recede once again. He has never felt so ill in his life as now.

The plan was to replace the canister on the scrubber with the final CO_2 canister only after they have been found, but the departure of *Pisces V* back to the surface, combined with their own fatigue, meant this wasn't done. The scrubber has been activated but is exhausted, and now it can do nothing more to clean the increasingly polluted atmosphere. To ration their remaining supply they have also delayed diluting the atmosphere with a fresh intake of oxygen. Every breath inhaled contains eight times more CO_2 than is considered safe. They have no choice, as pain now will later buy a few extra breaths.

The pain, however, is unevenly distributed. Mallinson is enduring far greater discomfort than Chapman, who while still afflicted by a fierce headache and fatigue, is well enough to recognise his good fortune in comparison with his partner, whose morale is in steep decline. There are moments when

Mallinson believes he will die, that he will never again see his wife Pam and his young children David, Michael and Vanessa, about whom he has begun to fret and worry. Chapman seeks to soothe his distress. He comes behind him and begins to cuddle up, as much for comfort as warmth, and then takes Mallinson's hand and begins to squeeze. As Mallinson later says of the comfort he derived from the touch, 'Words were not needed.'

The severity of Mallinson's response prompts Chapman to check the Ringrose CO_2 indicator, which states the CO_2 content in the atmosphere as 2 per cent, although he doubts its accuracy. It simply has to be higher. Among the emergency supplies is a small test tube designed to provide an alternative CO_2 test in the event of the failure of the Ringrose indicator. Chapman switches on the torch and begins a search. It's easily to hand, as it had been re-stored conveniently after the crash. He unplugs the airtight tube, allowing air from the atmosphere to gather inside the glass, and watches as it begins to discolour, matching the new colour to a chart showing rates of CO_2: the darker the colour, the higher the concentration. The result is disturbing. The colour corresponds to a new reading of 3.5 per cent, almost double what they had believed.

In spite of the new reading and Mallinson's deteriorating condition, they hold firm. The new canister for the scrubber will not be fitted until they are found.

At 11 am *Pisces V* begins her second descent and the rescue operation's third dive. When the men trapped below hear the news, Chapman tries to cheer Mallinson up by quipping that, hopefully, this will be 'third time lucky'. Yet his partner is in no mood for levity and Chapman's own mood begins to sour. As the countdown in metres begins again – 150 metres … 200 metres … 250 metres – he takes offence at the New World's metric system and Macdonald's soft Canadian drawl, and begins to silently fume at the absence of proper English feet and his fellow countrymen. Chapman is taking no comfort from strangers. He wants Des D'Arcy on the line. Anxiety is spilling over into anger as an irrational internal interrogation triggers questions to which he has no immediate answer: 'Why isn't one of the Oceanics team talking to us? Has *Voyager* left?'

In an agitated fit of self-conjured pique, Chapman violently turns over, knocking one of the two plastic timers off the ledge. He watches it tumble down to the 'floor' of the sphere, which before the crash was the stern wall. It's now a rubbish tip of damaged equipment and other detritus, stewing in a soup of inch-deep dirty water. To find the timer he needs to bend way down and fish around with his hands, yet as soon as he does so the smell strikes him like a sock to the nose. As carbon dioxide is heavier that oxygen and other gases that constitute the air in the atmosphere, it sinks, concentrating at the lowest level. Each time Chapman puts his head down to look, he's dunking into a poisoned atmosphere. Scanning the torchlight around the floor and feeling with his hands takes a while, longer than he hoped and, at times, longer than he thinks he can stand, but finally his fingers feel the familiar shape of the timer. Back on the bunk and now nursing an excruciating migraine, he uses a length of wire to bind the timer to his wrist. Worried that the drop damaged the bell, he sets the timer and is relieved when the world's ugliest wrist watch begins its irritating rasp.

On the seabed *Pisces V* begins again to search with sonar. The white submarine spins around in the darkness, her light casting out a few dozen feet, her sonar pings rippling out in expanding cones of sound for miles. The first sweep is unsuccessful.

Listening from inside *Pisces III*, Mallinson and Chapman cannot decide if their rescuers sound any closer than before.

Macdonald asks Chapman to again count to 50 over the underwater telephone. Chapman counts but it doesn't work. Macdonald wants to try something different. He asks Chapman to sing. The idea is that he might hit higher notes that have a better chance of penetrating through the water. He could have chosen from among the year's hit singles, which include 'Killing Me Softly with His Song' by Roberta Flack, 'Tie a Yellow Ribbon Round the Ole Oak Tree' by Tony Orlando and Dawn, or, on a more poignant note, 'Welcome Home' by Peters and Lee. He lacks the swagger to carry off 'I'm the Leader of the Gang' by Gary Glitter, but instead Chapman, not the most dedicated of pop music enthusiasts, ignores the successes of the American hit parade and *Top of the Pops*, opting instead for a composition of his own devising.

In his memoir Chapman suggested that the lyrics went something like:

Here we are, here we are
Somewhere near the cable,
Must be near the pinger
Depth is one thousand, five hundred and seventy-five feet
Come and find us
Come and find us

Mallinson begins to rouse himself but is an unsupportive backing singer, yet, like Chapman, he needs this song to be a hit. There's no melody, no recognisable tune, just words shouted in a lilting sing-song voice, but something in them catches Macdonald's attention, and as the submarine rotates he spots a stronger signal in one specific direction. Chapman, out of breath from his musical exertions, is preparing for an encore when Macdonald tells him he can step off the stage. They have a vague direction.

Pisces V has a sonar target at a maximum range of 1,600 feet, four times closer than on the previous dive, and heads in this direction, floating through the dark.

As the sub sets off, Mallinson listens intently to *Pisces V*'s sonar. He knows you can really only hear it if they're looking right at you. *Pisces V* is transmitting pulses, and these are going from one range to a second range as Macdonald adjusts the controls and the focus.

At one specific point Mallinson thinks he can detect a subtle change in the steady 'tick ... tick ... tick ...'. He believes, and Chapman agrees, that just as *Pisces V* changes over, the submarine is directly in front of them.

Mallinson takes over the phone and asks if Macdonald has just changed the sonar.

Macdonald replied that he has. Mallinson tells him that when he was changing over, 'You were looking exactly at us.'

This gives Macdonald the chance to put the sonar back to where it was, no longer sweeping like a windscreen wiper and passing them by, but more focused, with a tighter range and shorter sweep.

Mallinson thinks that the pair of them have found themselves on *Pisces V*'s sonar.

At a longer range, the space between the sonar clicks is more pronounced: 'click ... click ...'. But as the *Pisces V* gets closer, what they can hear is closer to: 'click-click-click'.

The prospect of discovery has a short-lived healing effect on Mallinson, who gets off the bunk, stands up and peers out of the porthole. The view is unchanged: a desolate sheet of black. Below on the bunk, Chapman dare not breathe, as if hope is a bubble a single breath can burst, yet he joins Mallinson at the viewing port. (In Macdonald's account, he asks Chapman and Mallinson to briefly switch on the exterior lights, then turn them off to best conserve the batteries.) All that both men can see is the impenetrable dark, but then something begins to happen. The dark starts diluting. What was previously a wash of black begins to slowly lighten in colour over the space of a few seconds, and then black becomes blue as their surroundings are bathed in *Pisces V*'s beams.

The men cannot see their rescuers. *Pisces V* is approaching beneath the level of their porthole, but they can see and will soon feel her presence. Mallinson's acute fears for his family renders him the more vulnerable of the pair to the emotion of the moment and he begins to cry, tears running down into his bearded cheeks. Hearing a sob, Chapman turns to him, and Mallinson, as if ashamed and not wishing to be seen, buries his face in Chapman's shoulder. It's a moment of intimacy and relief neither man will forget.

In *Pisces V* the shape on the sonar screen increases in size, and Macdonald and McBeth are confident they have them now. But it's not until their lights finally illuminate the scene that they relax and allow relief to briefly unknot the tension in their hunched shoulders. The scene before them explains why *Pisces III* has proved so elusive. The submarine did not land on a flat, open stretch of seabed, but inside a divot, a micro-trench that conceals a portion of the sub from sonar so only the top half is visible to detection.

'Roger, *Pisces III*, we see you now. Standby, we shall be manoeuvring around.'

On the surface Trice and Messervy break into broad smiles as they listen to *Pisces V*'s message to *Pisces III*. They have found them. The time is 12.44 pm. Messervy knows it should not take long to fit the rescue lines, and if all goes well this afternoon they should be home in time for tea. A message confirming the discovery is sent to David Mayo on *Hecate* with instructions to get it back to the base at Barrow. Trice and Messervy then listen in as Macdonald sends back a detailed report on what he can see: '*Pisces III* lying 80–85 to the vertical in soft mud, little exterior damage. After hatch opening visible.'

* * *

Pisces V glides gracefully around the static sub as Macdonald keeps up a running commentary and looks for any additional signs of exterior damage. Inside *Pisces III* Mallinson and Chapman prepare a toast. The solitary can of lemonade was to be their celebratory champagne, only to be opened when they'd been found. Chapman brings out the can, cracks open the top and passes it to Mallinson to take the first sip. As well as being in considerable pain, Mallinson is also parched. Dehydration is affecting both men, as apart from a few sips of cold coffee their liquid intake has been reduced to licking condensation from their fingers. The prospect of sweet pop is delightful, but when Mallinson takes a sip it's clear that particles of faecal matter from yesterday's unconventional toilet break have contaminated the can. It tastes like shit. Then Mallinson remembers why the can is open and decides that shit never tasted so sweet.

In the cloudy waters Macdonald adjusts the trim as he tries to position the sub as close as he can without actually colliding, though a couple of times the submarines kiss more than crash, their outer skins touching with just enough force that Chapman and Mallinson feel the tremor. Both boats are now at right angles to one another. *Pisces III* is standing almost erect, at just a little off 90°, with her stern burrowed into the sandy seabed and surrounded on one side by a natural sloping wall, like a man waist deep in a trench. *Pisces V* is floating seven feet off the bottom, her spotlights bathing her sister in white light that attracts halibut, cod and skate. Outside their shared cocoon of light is a dense blue blackness.

The manipulator arm holds the snap hook, which is tied in place by tethers. The snap hook is attached to the white polypropylene rope that floats in a loose billowing line, extending out and then up into the darkness above and for a further 1,575 feet to the surface, where on deck the crew watch its every twist and sway.

Macdonald looks through the porthole to where the manipulator arm bends to the touch of the joystick. Carefully he manoeuvres the arm up and over towards *Pisces III*'s steel lift point, which stands out against the frame like a white letter 'D'. The nose of *Pisces V* is a couple of feet back from *Pisces III*, giving the mechanical arm room to turn and extend. Although Macdonald has done this before, it doesn't make it easy, only familiar. The first attempt to push the snap hook inside the lift point fails, as it slaps against the outer edge. Macdonald pulls the arm back and tries again, but the angle isn't right and, again, the hook won't slip inside. The task is like threading a needle while encased in a suit of armour, but with patience, dextrous control and just the right amount of adjustment of the joystick, he knows it can be done. He stops for a second, then tries again, and this time the approach is dead on. The arm pushes out, and both Macdonald and McBeth can see it squeeze through the lift point's eye. The snap hook is through the 'D'. The needle is threaded.

For Macdonald and McBeth the intense concentration and low-level anxiety briefly dissolve in a cocktail of relief and delight. For the first time in over 50 hours a physical connection now exists between *Pisces III* and the surface: a rope strong enough to pull them from darkness to daylight.

Inside *Pisces III* Mallinson strains to see out of the port-hole, but because of *Pisces V*'s position, which is currently below the porthole's sightline, all he can see is illuminated water, brackish with the sand and sediment kicked up by *Pisces V*'s skids.

The delight of connection is pushed aside by McBeth, who spots that a coil of the lifting rope has fouled on one of their own external lights, so he eases the craft back, part of the plan to break away the loose ties that bind the snap hook to the manipulator arm, but which he thinks will also free the coil from the light. As *Pisces V* moves away, the coil unfolds and falls free and the loose ties break off, leaving the rope to drift up and the snap hook to catch tight on the lift point.

And then, slowly, the snap hook gently turns, like a key rotating in a lock, and falls out of the lift point.

Pisces III is again untethered.

Macdonald can't quite believe what he's seeing as the snap hook swings down in front of the observation port, floats for a second, and then, tugged by the floating rope to which it's attached, begins to drift up and away, like a balloon in a breeze.

The drift is slow, but once out of reach the snap lock and lift line will have to be reeled in. Macdonald and McBeth will then have to resurface and reattach, and then re-descend. That's two hours at least, three or more if they're unlucky. They have to try to catch it. McBeth quickly pivots *Pisces V*, sending the stern slamming into *Pisces III*, a body blow that almost knocks Mallinson off his feet and startles Chapman, who has no idea what has happened.

Macdonald grabs the joystick and, as *Pisces V* turns, he reaches out the manipulator arm, which moves at the pace of a pensioner, almost but not quite as slowly as the drifting line. The difference between the arm and its quarry, the fleeing hook, is dividing down: four feet, two feet, one feet, six inches, three inches then, just as it's about to slip away beyond the arm's reach, the hook catches on the edge of *Pisces III*'s light bracket and – briefly – is held suspended, despite the weighty tug of 2,000 feet of line.

The manipulator's claw hand gently reaches up and clasps the snap hook. Once it's secure, Macdonald spins the arm around, looking for somewhere to temporarily secure it. The closest point to hand is the steel grille that covers the propeller. He slams in the snap hook, which locks onto the grille. They have a connection, but it's useless.

CHAPTER ELEVEN

The antidote to bitter disappointment is distraction. Chapman and Mallinson decide it's time – finally – to replace the lithium hydroxide canister on the scrubber. Mallinson is still suffering from exhaustion, fatigue, joint pains and a relentless headache, and Chapman can feel himself slipping down a similar path towards incapacity. Replacing the canister would take five minutes on the surface, but it takes the combined efforts of both men almost an hour as they struggle to unscrew the existing canister and fit the replacement, which is their last.

By the time they're done with it they're sweating and panting out more carbon dioxide, but at least now they can lie down and rest, with the scrubber working efficiently for the first time in several hours. As the carbon dioxide level is slowly scrubbed out of the atmosphere, Chapman bleeds in more oxygen from the remaining tank.

When Chapman checks the oxygen supply of this, their second and final oxygen tank, the reading is 1,800 psi, the equivalent at their current, highly restrictive use of less than 24 hours' supply. They have been submerged for 60 hours and it's now 51 hours since the accident. *Pisces V* has found them but a rescue line has not yet been attached, and it's clear to Chapman that they have to be on the surface with the hatch open to the air by 12 noon tomorrow.

Chapman explains the new time frame to Mallinson. Neither man needs to discuss the consequences of failure.

At least the batteries are holding up. Earlier concern about the level dropping to 100 volts has diminished as the power supply remains steady.

On *Voyager*, delight at the discovery of *Pisces III* begins to dissipate as garbled messages begin to drip through from *Pisces V* about a 'difficulty' with the lift line. Into the brief information vacuum Messervy, Trice and Bob Eastaugh pump in different scenarios as to what might have gone wrong. They know that this time, buoyancy of the rope isn't the issue. Perhaps *Pisces V* has accidentally struck *Pisces III* during the manoeuvre and lost the line? Then Al Witcombe picks up part of a message that the lift line is not connected to *Pisces III*'s central lift point but to the submarine's starboard propeller grille. Everyone in the room knows this will never support the weight of *Pisces III*.

Something has clearly gone wrong, and when communication improves they are initially quite stunned to learn what has happened. The snap hook was in and on, yet somehow just fell out? Trice had argued for the use of the snap hook over Vickers Oceanics' specially devised hook. He now falls quiet. When he needs to think he likes to puff on his pipe, and this would be a particularly good time to fire up the briar.

Pisces V is told to stay with *Pisces III* and once again attempt to connect the snap hook to the lifting point.

Pisces V floats seven feet off the seabed and a few feet distant from *Pisces III*. McBeth works the controls and tries to move the snap hook from the propeller grille back up to the lift line. It won't work. He tries a number of times over the next 90 minutes, but he can't get it to fit, so he places it back on the grille. The battery gauge indicates that the battery is draining down. All the criss-crossing of the seabed in search of red herrings and submarine-shaped rocks has taken its toll. If they want to have enough power left to surface, they can't continue like this. When they report the battery situation to *Voyager* it's in the expectation that they will be told to come back up and that *Pisces II* will take their place. Instead they are told to hold their position.

On *Voyager* Messervey and Trice examine their options. The *John Cabot* has left Cork but is still a few hours distant; *Pisces II* remains a work in progress and the technicians' latest report is disappointing. They had hoped it would be an easy repair, but instead it needs to be stripped right down – almost a complete rebuild. All attempts to secure a definitive completion time result in waffle and an angry insistence from the men that they are working as fast as they can.

If *Pisces V* can't get the snap hook onto the lift line, is there another way to get a second line down, one more easily manipulated that could connect to the lift line? A sketch is drawn up of a 'choker', a piece of metal and rope 40 inches long, looking a bit like a pendant, that slides down the polypropylene rope from the surface to *Pisces III*. One end would be secured to the rope, and when it landed on top of *Pisces III* the remaining loose end could be picked up by the manipulator arm and run down to connect to the principal lift point. The plan is approved and construction begins.

By 2.30 pm *Pisces V* is almost out of battery power and lies inert on the seabed, her spotlights extinguished and the manipulator arm motionless. On board Macdonald and McBeth are weary after eight hours of intense focus and no little stress, and are stretched out on the bunk, drifting into sleep. For now, there's nothing else they can do. A dozen or so feet away, across a short but unbridgeable expanse of ocean, Chapman grows increasingly worried about Mallinson's condition. He seems to be deteriorating. The headaches are worse, and so is the pain in his limbs. The scrubber is now working as it should, and Chapman hopes the rise in air quality and fall in carbon dioxide will eventually ease his distress, but as yet it has not. In time he will update the surface that 'Roger's condition has deteriorated slightly and I am a little concerned. However he is bearing up well.' It's as if he wants to indicate a bad situation has gotten worse, while maintaining the British abhorrence of making a fuss.

The message is picked up by the press in the neighbouring fishing boats, among whom is Simon Dring, the BBC correspondent on board the *Marina*. The report he dispatches indicates that both men's condition is deteriorating drastically, and this will soon lead radio bulletins back in Britain.

* * *

Sir Leonard Redshaw spends most of the morning and early afternoon in the Portakabins of Vickers Oceanics and their temporary command centre. Like everyone, he's both frustrated and disappointed first by the buoyancy problems with *Pisces II*'s lift line, then by *Pisces V*'s failure to secure the snap hook. An additional frustration remains the poor communications between Barrow and the accident site. There's no direct link to *Voyager*, with calls still having to be booked in advance and connections inconsistent. Messages are coming via the Ministry of Defence and HMS *Hecate* and arrive second- or third-hand, with embellishments and misunderstandings embroidered in along the route. Hopes of using the communication system on board the nuclear submarine in Barrow's dry dock were extinguished when the system was defeated by the local topography. When Redshaw learns about the behaviour of the media he is enraged and decides to address the issue at the next press conference.

Many staff have brought in portable radios. Barely a bench or desk is without one as they listen attentively to each news bulletin. The message that arrives back in Barrow is understandably devoid of nuance. Irrespective of who hears it and who tells it, everyone knows that the men are in serious, life-threatening trouble. They are, but this is because of the imminent exhaustion of the oxygen in less than 24 hours, not, as it's portrayed on the radio, the current build-up of carbon dioxide.

* * *

They are running low on rope on board *Voyager*. Giant coils curl high up on deck but the rescue operation needs new supplies of a lighter line, and Eastaugh and Trice want it by nightfall. After calling around the neighbouring vessels, a supply of 1,800 feet of 1.5-inch lightweight braided nylon line is sourced from HMS *Aeolus*, whose crew arrange for it to be transferred across. There may also have been an attempt to arrange a ship-to-ship transfer using a rope sling. Al Witcombe sees a rocket line fired over to *Voyager* snag on the bridge just outside the window where he was manning the underwater telephone, but when he runs out to catch it, *Aeolus* had already began to pull away and the rope tears through his hands, leaving a rope burn as it whips out of reach.

The rope is instead to be delivered by helicopter. Eastaugh and Trice are on the bridge watching as the chopper takes off from *Aeolus* with the rope coiled in a net-like basket secured by a rope to its underbelly. It's 6 pm and the early-evening light casts a pale wash through the rising wind and rain into which the helicopter begins to rise, the line underneath tightening and the basket lifting from the deck as it swings out over the ocean.

Then the helicopter begins to lose altitude. Trice and Eastaugh can see it heading towards them but dropping down fast. At the controls the pilot is struggling to gain altitude as it becomes clear the total weight load of the rope has been underestimated. The basket is dragging the helicopter down and it's seconds away from crashing. The pilot decouples the load.

Watching from the bridge through the panoramic screen of the window, Trice sees the helicopter move closer to the water even as the basket detaches and tumbles down, splashing into the sea. The aircraft recovers and banks away, leaving the lift line bobbing among the waves. Trice has a saying during such trying times: 'If it was easy, everyone would do it.' Nothing about this rescue has so far been easy, and as they approach sunset, it seems unlikely that anything will be easy in the immediate future, certainly not in the dark.

By 7 pm a second supply of 1,800 feet of lightweight braided nylon rope is successfully delivered down onto the deck of *Voyager*.

The choker is ready and carried over to where the lift line, currently connected to the propeller grille of *Pisces III*, rises up and connects to the ship. One end of the choker is connected to the line and the other left to hang free. This is the end that is to be connected to the lift point by *Pisces V* using the manipulator arm. The choker is tossed overboard and, like a man with one hand on a banister of a spiral staircase, begins its circuitous journey to the bottom of the sea.

The *John Cabot* departs Cork at 6.45 am. So extensive are the supplies brought over by the Americans that moving them by barge upstream to where the ship lies at anchor takes longer than initially envisaged. (In fact, so replete is their inventory that supplies are deliberately abandoned on the dock, discounted as unnecessary, prompting an anxious call from the Vickers Oceanics temporary office by the docks to inquire if the equipment has been left by accident.) After an 11-hour sailing, the ship arrives at the site at 5.30 pm.

The American cavalry have arrived, and the first task requested of the captain is to quell the rebellious interlopers. The fishing boats with press on board still refuse to push back and continue to encroach on the accident site. Messervy wants the *John Cabot* to police the boats, using her size to intimidate them into backing off, then secure their camp-fire corral.

June Chapman has spent the past 48 hours in the company of George Henson, who insists she stay with him and his wife Marjorie at their traditional Cumbrian farmhouse in the Cartmel valley. Lovingly renovated, quaint and comfortable, with well-tended gardens, the home provides a warm welcome, the evening meal a pleasant distraction from the stress of the day. June returns to the office every morning in an attempt to distract herself with routine office work, although she joins Henson for every phone call, sharing the receiver so they can both listen to any updates. Henson believes, and June agrees, that it's vital to a joint coping strategy that during such phone calls – and any meetings down at Vickers Oceanics – that no emotion is shown.

So on Friday afternoon, when Greg Mott calls Henson from the operations room at Vickers Oceanics, June steels herself. Mott says they are receiving reports that the condition of both men has deteriorated and that they have become delirious. Mott adds that he hasn't yet been able to verify the reports, suggesting that they may have come from radio news bulletins rather than from either a call to *Voyager* or a call from their temporary office in Cork. Mott says he will ring back.

To June, listening on the line, Mott's voice sounded as grim as the picture he painted. Roger, a man of rational restraint, incapacitated and thrashing around delirious is too awful to imagine. Yet she can readily visualise his plight; earlier, on a visit to Vickers Oceanics, she had been shown inside *Pisces I*. June had wanted to know the environment in which Roger was living and, in her darkest thoughts, where he could die. She also wanted to know the arrangements for the toilet, which, although she knew it was a minor concern compared with their constant exposure to danger, had greatly bothered her.

It took some time before the matter of the men's health was clarified. The media, working with limited information, were unrestrained in their speculation. Mallinson's health problems became the plight of both men. Mott may also have overreacted, as when he asked a member of staff to prepare Pamela Mallinson for her husband's death.

The Cobb, the long, curved stone wall that arches out like a protective arm from Lyme Regis on the Dorset coast, forms a safe harbour that has secured the town's fortunes for centuries. On Friday afternoon, in a small aquarium housed in a stone building dating to 1723 at the Cobb's end, Ralph Chapman is drowning his worries in work. Since retiring from the armed services – the navy during the Second World War, followed by a stint in the army in the Far East – Chapman senior has helped run the town's local aquarium. Opened in the 1950s, the building is a simple two-room structure that in centuries past served as an isolation hospital for sailors stricken with cholera, typhus and, on occasion, the plague. The job is part-time and involves showing schoolchildren how to hand-feed the mullets in the water tanks and the right way to gently handle a starfish, with a little local history stirred in along the way.

On this stormy, windswept afternoon the waves are crashing hard against the stone walls, the air all salt water and sea spray. There's an atmosphere of foreboding, one captured four years earlier by the town's local author John Fowles in his bestseller *The French Lieutenant's Woman*. Ralph Chapman is suffering the anxiety of a father whose son is lost in the ocean depths, and every glance out the window at the cold grey sea is a bitter reminder. He has brought down his portable transistor radio to listen to the news bulletins for updates on the rescue.

In the family house up on the hill, his wife Hilda receives a phone call from staff at Vickers Oceanics urging her not to listen to the news reports on the health of her son and Roger Mallinson. The information, the caller explains, is incorrect and he will call back shortly with an update. Hilda worries that Ralph will hear the erroneous broadcast on his portable transistor and, wanting to reassure him but unable to drive, she sets out on foot into the wind and rain. Hilda descends the steep path from the couple's home towards the sea front and the Cobb, and when she reaches the Cobb's stone walkway she battles along its length till she reaches the entrance to the aquarium.

Ralph is sitting at the reception desk. He's surprised to suddenly see his wife, fearing the worst, but she explains about the call from Vickers Oceanics and quickly tells him that their son is not as badly affected as the radio bulletins have reported.

After reassuring Ralph, Hilda heads out into the driving wind and rain, her head tucked down, but instead of walking straight back along the Cobb's main arm, she turns right to where the harbour wall extends for a short distance then ends at the sea. She is lost in thought and unaware of her mistake. Ralph only belatedly realises which way she has turned. He sprints out of the door, sees her lonely figure, head still down, and manages to catch her just before the last fatal steps.

Over the course of the day Ted Carter, the assistant technical manager at Vickers Oceanics, has grown increasingly worried about morale on board *Voyager*. When he speaks briefly on booked calls he thinks he can detect a growing despondency following all the setbacks, and he knows for a fact that no one on the vessel has slept in over 50 hours. He also feels that problems are mounting and there's more that can be done.

This morning he's got word back from Oldham Batteries, whom he contacted to check about the effect on the sub's batteries from the near 90° angle at which the vessel is standing. The company's research has concluded – two days later – that the critical angle for electrolytic leakage is 90°. According to their report, the batteries should lose 50 per cent of electrolyte. Carter knows this hasn't happened yet, that according to the reports from *Pisces III* the batteries are down slightly but holding steady. Yet the fact is that this could change at any moment. No battery, no scrubber. No scrubber, and the carbon dioxide builds up. It's another turn of the screw.

Carter decides the rescue operation requires a back-up team, an injection of fresh blood and expertise undiluted by draining days in the wind and rain. As Sir Leonard is in the office at Vickers Oceanics for an update, Carter approaches him. In normal circumstances this would never occur. Redshaw is like a god, a towering figure in control of over 10,000 men, but today he's just like all the rest of them, a man worried about the fate of his fellow workers. Yet Redshaw is different in one crucial way: he can get things done.

Carter says he wants to put together a new team: pilot, engineer, acoustics, diver, etc., and get out to *Voyager* as soon as possible. Redshaw agrees and tells Carter to get what he needs and that he will speak to the Ministry of Defence about getting them on board tonight. The only viable way is by military helicopter.

In the third-floor office of VDPT, June Chapman is working late with George Henson, who liked to back up administrative paperwork to the very end of the week, where it had nowhere else to go. While others would resent their boss's eccentric clerical habits eating into their weekend, June doesn't mind, and tonight they make a welcome distraction. The work is complete by eight o'clock, and both pay a last visit to Vickers Oceanics. The visit is brief as there are no new developments, simply tired men waiting for word from the rescue and reluctant to leave without it.

The atmosphere between Henson and June is sombre and largely silent. Even when they visit a small restaurant for a late supper they find there's little to say beyond routine platitudes about the quality of the food. Tonight June has decided to forgo the company of the Henson household and will return to the cottage in Broughton-in-Furness. En route home, they stop at the house of Agnes Steele, or Jinny as she's known to all around the village. Almost a second mum to the young couple since they moved in earlier in the summer, Jinny is pleased to see June but her distress at Roger's plight comes as a shock. The look of strain and worry etched on Jinny's face briefly startles June, who, lost in her own bubble of worry and fear, has forgotten that the accident might be affecting others. Jinny is so worried that the expected roles are reversed. June now mothers her, insisting that all will be well and Roger will be rescued tomorrow. June speaks the words, but doesn't know if she believes them.

As afternoon rolls into evening Chapman sleeps while Mallinson, whose condition has improved slightly with the rise in oxygen and drop in carbon dioxide, operates the scrubber. Both men are fast asleep shortly after 5 pm, when the sphere is filled with an excruciating whine, like guitar feedback from the world's most powerful speakers. A few feet away, Macdonald and McBeth in *Pisces V* are responsible for the noise, having tested their pinger at close quarters. At 5.15 their sub powers up, switches on her exterior lights and rises from the bottom, illuminating the upturned body of *Pisces III* as they move past, until settling into a gentle hover by the propeller grille.

The voltage on the battery inside *Pisces V* should be at 120 but has dropped to 67 volts. The pair had been told that the choker had already been sent down the line, but after waiting for its arrival Macdonald and McBeth had given up, assuming it had become trapped by a kink on the line. As they waited, they took turns to climb into the sleeping bags they'd brought to keep them warm and get some rest.

Two hours after the choker was thrown into the water, Macdonald and McBeth are woken by an audible clunk coming from the side of *Pisces III*. What they refer to as 'the pendant' has touched down. Their battery power is so low that they can only switch on the exterior lights for five seconds at a time. During this snapshot they peer out of the porthole and can see the thick grey rope, fixed by the snap hook to the propeller's steel guard grille. The rope rises up towards the surface. At the bottom of this rope is the newly arrived choker: a length of rope attached by a steel latch to the line at one end, while the other end is free and fixed with a steel hook.

Their task is to grab the hook with the manipulator arm, drag it across to *Pisces III*'s central lift point then, once again, thread the needle. If achieved, there will be an uninterrupted secure line from the central lift point to the surface. One tug on the propeller grille and it would snap clean off. One tug on the lift point and it will hold, and hopefully carry them to the surface.

The difference with 'threading the needle' this time is that it has to be performed while half blind. They can't risk dividing the remaining battery power between the external lights and the manipulator arm – it's one or the other. So to illuminate the operation Macdonald switches on the small hand-held torch, holds it up to the porthole window and tries to light up the propeller grille and choker. The porthole glass reflects back some of the light but, lying beside him at the controls, McBeth can just about see what he's doing, the torchlight moving back and forth on the choker's free end. The manipulator arm reaches out and the claw gently grasps the hook, then drags it through a cone of torch-lit water in an ocean of black.

Then the battery runs down. Everything is left frozen and floating. The battery needs to recharge and so they must wait for what little power they have used to recoup. For the next two hours, attempts to fit the choker's hook into the lift line's eye will be sliced into incrementally smaller moves that seem to leave them, paradoxically, both closer and ever further away from the target.

Time drains away alongside the battery. Finally, both men realise the task is impossible under such restricted power supply, and *Pisces V* sinks like a beaten boxer onto the rippled seabed.

Inside *Pisces III* Mallinson and Chapman lie in an exhausted semi-stupor. Hours drift by in either sleep or semi-consciousness, a routine broken by the activation of the scrubber. The atmosphere is more fetid than ever. In the afternoon Chapman, like Mallinson before him, could hold his bowels no longer and prepared to defecate into another plastic bag, but unfortunately exhaustion, the creeping cold and an inability to concentrate, meant the bag was incorrectly positioned to catch his waste, which instead plopped into the water at his feet. Furious with himself for adding to their discomfort, Chapman tried to fish around by hand to retrieve his turd, with a view to bagging it then sealing it securely in the tin box alongside Mallinson's waste. But it had already floated out of sight, and after a few futile moments he and Mallinson decide the energy and excess air expended in the search were disproportionate to the reward.

They can endure one more shitty night. One way or another, both know it will be the last.

Now that all the players are on the scene, Messervy calls a conference to plan the next move. The *John Cabot* has completed a circuit of the accident site and made an effort to push back the press boats. Afterwards a helicopter is sent from *Aeolus* to the *John Cabot* to pick up the American team. One by one Lawrence, Watts and Commander Moss are winched off the deck, flown half a mile to *Voyager*, then winched down. It's 7.30 pm and the winds are beginning to blow with increasing strength. In an hour or so it will be skirting a storm.

In *Voyager*'s galley the strict rules laid down by Captain Edwards have been relaxed on Messervy's orders. Yesterday Al Witcombe and the Canadian contingent had arrived in oil- and work-stained clothes, weary from preparing *Pisces II*, only to be refused entry on grounds of ill dress. Witcombe was too tired to care and took the attitude 'to hell with your food'. When Messervy found out, the steward was ordered to accommodate everyone, regardless of how oil-stained their clothes. Afterwards George, the chief cook, never let a man pass the door without hollering out, 'Hey, friend, want a bacon sandwich or steak and eggs?'

Tonight the menu is coffee for the Americans and tea for Messervy. They are joined by Trice, who listens, puffing softly on his pipe, while the current position is laid out. The initial estimate was that oxygen supplies on *Pisces III* would be exhausted by 10 am the following morning, less than 15 hours away. Yet the most recent update from Chapman is that the submarine's oxygen supply will stretch to 12 noon. *Voyager* has been at the accident site since 1 am, and in the past 18 hours has managed only to locate the sub and fit a lift line to the propeller grille, a purchase of inconsequential use. Messervy explains to the Americans the problems they encountered with the buoyancy of the original rope, the torn manipulator arm on *Pisces II* and the issues with *Pisces V*. Lawrence, Watts and Commander Moss listen but get the sense that the captain still wishes to keep them on the subs' bench: ready in reserve, but hopefully never to take to the field.

Messervy is pleased to say that *Pisces II* is almost ready to launch. The repairs to the manipulator arm are complete and Des D'Arcy is, even as he speaks, completing the final check list. The plan is for an 8 pm dive that will see *Pisces II* home in on the active pinger on board *Pisces V*. Once they are in situ, the team will fit the toggle straight into *Pisces III*'s aft sphere. If all goes well, the men could be back on *Voyager* by midnight.

There are two toggles on board *Voyager*: the one that is now bound to the manipulator arm of *Pisces II*, and a spare that Messervy wants the Americans to take with them back to the *John Cabot* and CURV-III.

*   *   *

Two decks below where Messervy is holding court, the braided nylon line brought over by *Aeolus*, on the second attempt, is being fixed to the manipulator arm, and for extra stability it has been taped along the hull. At 7.50 pm *Pisces II*, with D'Arcy and Browne at the controls, is lowered over the side into what the *John Cabot* log describes as 'rough sea and heavy swell'. At night, waves can seem so much bigger, and Bob Eastaugh, who is on the stern of *Voyager* supervising the launch, already estimates them at between 15 and 20 feet. Under no other circumstances would a launch proceed in such conditions.

The metal cradle disconnects and the divers scramble from the rocking Gemini onto the top of *Pisces II* to connect the tow line. Lights from *Voyager* illuminate the scene, but beyond their glow is a deepening darkness. When the strop is properly secure, the diver instructs the Gemini pilot to begin to move, then rides out straddling the sub's fin and bounces through the waves till they reach the white metal buoy. Once in position, the diver disconnects the tow line, and over the VHF radio Eastaugh gives D'Arcy and *Pisces II* permission to descend.

Inside *Pisces II* D'Arcy has only just begun the descent, watching through the portholes as the vessel slides down through the waves and moves below the surface. The submarine is barely below when the sphere is suddenly filled with the incessant demand of an emergency alarm. D'Arcy knows almost immediately that it's the water alarm, designed to be activated when the aft sphere begins to flood. Can this really be happening? They are on a rescue mission to retrieve a sister sub whose aft sphere flooded, sending them nearly

1,600 feet to the bottom of the Atlantic, and now the alarm shrieks that exactly the same thing might be happening to them. D'Arcy knows the water alarm is super-sensitive and that it could have been triggered by damp or the smallest droplets of water, but he also knows he can't take a chance. They have no choice. They have to surface.

Eastaugh looks out through the rain to the floodlit spot where the sub slipped below the surface on his command and is startled to see her suddenly reappear. 'What the fuck is going on?' he must be thinking. D'Arcy hasn't yet communicated to the surface crew what has happened, but Eastaugh recognises that he has an emergency on his hands and shouts to the crew to prepare the 'mine' bag. This is a giant rubber sack that can be inflated by canisters of compressed air and will support the weight of a waterlogged submarine. It's used by the navy to support unexploded mines, leftovers from the Second World War that continue to bob about in the Atlantic almost 30 years on from the conflict's end.

At sea, the diver in the Gemini has his view obscured by the crests of rising waves. As the wave breaks he spots the submarine in the wave trough and heads back towards her. The waves are so high that for seconds at a time the diver and Gemini pilot lose sight of the sub, and when she reappears it briefly looks as if the stern is sinking down lower than the bow. From the deck of *Voyager* Eastaugh thinks he can see the same thing: a repeat of what happened on Wednesday morning. Has the aft sphere flooded again? Are they going to have two crippled submarines on the bottom?

The mine bag is hurled into the water, where a second Gemini craft with pilot and diver battle the weather and waves to transfer it out to where *Pisces II* has resurfaced. By now D'Arcy has used the VHF radio to inform *Voyager* that the water alarm has been activated but, as far as he can tell, the vessel is currently stable and they don't seem to have taken on any extra weight from an ingress of water. D'Arcy and Browne have braced themselves for an accident that hasn't yet come to pass, and are now dealing with the disappointment and frustration of another rescue attempt ending before it has even begun. Outside, the divers decide the mine bag doesn't need to be activated as the sub is sitting stable in the water, so they connect the tow line and begin to drag *Pisces II* back to *Voyager*.

The recovery takes 25 minutes and by 8.30 pm *Pisces II* is back on deck.

Worst-case scenario: they have 13½ hours left. Best-case scenario: they have 15½ hours left. They have been on the job 17½ hours with not a single workable lift line secure.

Any hope Messervy may have entertained of displaying Vickers Oceanics' nautical prowess to the Americans – of having the 'Yanks' on board when the first lift line is secured and, in doing so, re-asserting his command of the rescue operation – is now shot. Instead Lawrence, Watts and Moss have watched as the third submarine attempt has ended in failure. Messervy has, for the time being, no more pieces to play. *Pisces V* is still on the seabed with limited power; *Pisces II* has only just been recovered, and is the subject of an in-depth inspection to find the cause of the alarm sounding and whether there are any other undiagnosed technical problems.

It's time for the Americans to make their move. To signal the passing of the baton, Watts and Moss take possession of the second rescue toggle. The question is how to get both men and the toggle, which on its own weighs 125 lbs, back to the *John Cabot*. Darkness is now descending and the helicopter on board *Aeolus* is not only low on fuel but also, frustratingly, unequipped for a night flight. Instead the men are going to have to take the toggle on a Gemini, ride those dark and mountainous waves back to the *John Cabot* and tell the team that CURV-III has been given the go-ahead. The opinion of Moss and Watts is that CURV-III will be ready to dive in less than one hour.

Bob Watts wraps the toggle in two lifejackets and lashes it with rope to the Gemini in case it falls overboard. Before the boat enters the water, Watts requests that Captain Edwards contact the captain of the *John Cabot* and ask him to turn the vessel towards them with spotlights on and either a scramble net or a Jacob's ladder at the ready. The journey from *Voyager* to the *John Cabot* is only half a mile or so, but the sea is treacherous and the waves are hitting 25 feet. To Watts they resemble 'giant black mountains'. In each wave's trench neither ship is visible, only dark water, while the noise of the wind, which Watts estimate is blowing 50 knots, means both men can barely hear each other shout.

As they approach the *John Cabot* it's clear that Edwards hasn't called across, or if he has, the captain hasn't been told. There are no spotlights, no scramble net or ladder, and no one on the lookout for them. Watts shouts up, but it's clear that he'll never be heard, not in this weather. In the darkness the Gemini begins to batter against the ship's hull. Then they have a stroke of luck. One of the galley crew is tossing kitchen waste over the hand rail when he catches sight of the two men below. A minute or two later, spotlights illuminate them and a Jacob's ladder is quickly lowered down.

In the wood-panelled conference room at Vickers, the press conference on Friday evening is a dispiriting affair, taking place in a fug of cigarette smoke. Sir Leonard Redshaw, in a sombre grey suit, white shirt with silver cufflinks and dark, lightly spotted tie, keeps his arms crossed, both holding himself together and panic at bay. Mott, the managing director, his hair in a lightly oiled quiff, leans forward with his hands clasped in his lap. Neither man is able to fully express the peril of what's occurring far away in the Atlantic.

Redshaw kicks off by outlining the situation: 'I would not like to calculate odds. Now it will be a neck-and-neck thing if they are brought up. We have trained ourselves not to get too optimistic. I must admit at the moment we are a bit down … we knew time was against us, although we had 72 hours to play with. Time is against us. There's no denying that we would like another 24 hours, but this is always the problem. Time is not on our side.'

The company also makes clear their displeasure with the BBC and their reporter Simon Dring. In a statement that presents their complaint as a united front, the naval support ship *Hecate*, the Commanding Officer Plymouth, the Ministry of Defence Operations Room and Vickers all state that they 'strongly objected' to Mr Dring's activities. 'In spite of this, Dring continues to issue distorted information being broadcast on the BBC, and it's causing unnecessary panic and extra effort by our hard-pressed team.'

The TV channels BBC1 and BBC2 will not carry reports from Mr Dring on their 11 pm bulletins, but the broadcasting corporation denies any charges of interference or improper reporting by its man in the Atlantic. In a statement released in response to Vickers' official complaint, the BBC insists that there was no interference with radio communications and no inaccurate information given. The BBC press office responds: 'Mr Dring's information is derived solely from *Voyager*, the mother ship. He has passed that information on, where he was able, to the listening and viewing public. Our researches do nothing to change our opinion on this matter, but we fully appreciate the great pressure under which everyone is working.' But the BBC is not the only the target for complaints. A spokesman will also take aim at the fishing trawlers hired by newspapers: 'The trawlers in the area are getting in everyone's way.'

Redshaw talks reporters through the problems they have encountered during the day, but talks up the physical condition of Chapman and Mallinson, saying they are fit, although he fears the stress they are under. 'They must be under tremendous pressure by now. We have been keeping them informed of every development and hitch in the operation. But I suppose if something drastic happened towards the end of the rescue we would keep it from them, because the thing these men need most of all now is hope.'

After the press conference, Mott slips away to a quiet office to call Maurice Byham, who is still looking after Pamela Mallinson and the couple's children at a friend's house in Windermere. Mott feels it's now appropriate to tell Pamela that there's a strong likelihood that her husband may not be rescued in time. He tells Byham that he has to tell her, he has to let her prepare for the worst. Byham listens, agrees, then hangs up. But he's telling her nothing of the sort. He believes where there's still air and time, there's still hope.

When asked about the submarine's air supply, Redshaw chooses the most optimistic answer: 'We estimate that the men's air supply is good until noon tomorrow.' Before the press conference is drawn to a close and the reporters pick up their shorthand notebooks and head to the nearest public telephone, followed by the pub, Redshaw makes a statement that captures the mood of everyone in the office and at sea. 'We've had blow after blow. We really do believe we deserve a break.'

PART V

SATURDAY

CHAPTER TWELVE

The helicopter in its wasp-yellow livery is waiting on the tarmac of Cork airport when the Vickers plane touches down just after eight o'clock on what is an increasingly stormy Irish evening. The fading light of late summer has bruised the sky into a palette of russet red, grey and gold and, as is often the case on the west coast of Ireland, there's the promise of heavy rain.

The Royal Navy long-range Sea King from 824 Squadron has taken off an hour earlier from RNAS Culdrose in Cornwall, after receiving final clearance from the Irish government. They won't tolerate British military aircraft in their airspace, especially the frequent, illegal incursions in the north that have become almost insultingly routine as 'The Troubles' expand, but, for now, everyone in London and Dublin is pulling together. In preparation for a prompt departure, the pilot starts the rotor blades when he sees the Vickers plane's wheels touch down on the landing strip.

Ted Carter and his four-man team disembark and hurry across the runway, where they're joined by the newest member of the relief team, a local doctor. Vickers has taken almost two days to confirm that among all the crew on board what is now a growing flotilla at the accident site, there's currently no medical provision beyond a well-stocked first-aid box. The men are briefed on the upcoming flight by the air crew, strapped into their seats and handed ear protectors to quieten the deafening din of the rotor blades and roar of the Rolls-Royce Gnome turboshaft engine. The Sea King, adapted from the American Sikorsky, has an operational range of 600 nautical miles and a top speed of 150 miles per hour, and she is not quiet.

A few minutes later the helicopter lifts off above the airport, then arcs out over the encircling tapestry of green and brown fields towards the coastline, and on out to the big blue of the ocean, into what they've been advised are strong headwinds and a gathering storm. On board there's a palpable sense of excitement, mixed with fear. None of the five men from Vickers have flown in a helicopter before, much less with the prospect of being winched down onto the deck of a rolling ship in the teeth of a wild Atlantic storm. Ted Carter looks down at his faded blue jeans, polished dress shoes, checked flannel shirt and worn tweed jacket, and thinks, 'At least we're dressed for it.'

After more than an hour's flight through increasingly foul weather, the pilot spots the flotilla on the horizon. As the helicopter approaches they can see the grey hulk of *Voyager*, rolling in heavy seas, and 400 metres to the stern the low, flat features of the *John Cabot* and HMS *Hecate*, around which is dotted a posse of fishing boats, their lights twinkling in the gathering gloaming. The pilot's concentration is fixed on the destination, but Dick Nesbitt peers out the window and can't quite believe what he is looking at:

I can remember looking down from the Sea King helicopter as it hovered above the *Voyager*. It was at night and very dark. The ship was all lit up and when I was looking down the sea – the wind was so severe, I think it was Force 10, but very, very severe weather conditions – the wind was blowing the white surf of the sea and it looked like a polar landscape. I thought: 'Is that water or is it that snow?' It was just the 'white horses' of the waves and the tops had been blow across by the wind.

At 9.30 pm the Sea King moves into a steady hover directly over *Voyager*. The helicopter's heavy steel door is slid back and the men are greeted by a vertigo-inducing view of a 100-foot drop down onto the cleared cargo deck. The winchman draws in the steel cable line and prepares the cushioned sling, then slips it over the head of Ted Carter and fixes him into the harness. As de facto team leader, Carter is first out of the door into what he describes as 'a risky, risky experience'.

When he's swung out by the winchman, there's only air under his shoes and the wind is battering his face, making it tricky to breathe. There's no noise other than the deafening din of the rotor blades whirring just 10 feet above his head. Below he can make out the white lights of *Voyager*'s deck and the outstretched arms of the crewmen, while on the horizon is nothing but the white tops of the ocean's cresting waves. Carter is steadily winched down till his head is level with the winchman's boots, then he's looking at the helicopter's under-belly. In the cockpit the pilot is keeping the aircraft level with the pitch and roll of *Voyager*, but for Carter the problem comes when he's within a couple of feet of the deck. The crew grasp at his legs, but as soon as his feet touch down, the deck falls out from under him and he's left suspended 20 feet in the air as the ship plummets into the trough of a wave.

The winch operator tries a second time, then a third. During the fourth attempt the ship is buffeted and knocked starboard, leaving Carter's legs straddling the railing with one foot over the Atlantic. Then when his feet do touch the deck, there's a race to get the harness up over his shoulders and clear of his head before the ship tumbles down the next wave trough. Carter finally escapes the harness, and he's hurried into the aft accommodation block. Back out in the wind and heavy seas, the winch transfer of the rest of the team is equally problematic, with men clattering down onto the steel deck then being hauled back up into the sky. The doctor who came to treat any potential casualties almost becomes his own first patient when he lands severely.

Messervy is watching the helicopter transfer from the bridge, and as soon as he receives confirmation that Carter is on the deck he heads down to see him. Messervy is actually the first person Carter talks to when they meet in the stairway of the aft accommodation block: 'To be honest, he was really, well, he hadn't given up, but he was more or less in tears.' Messervy gave Carter the broad strokes. CURV-III, the American remote-controlled vehicle, is preparing to launch in the next hour or so. *Pisces II* has just been recovered after the water alarm and *Pisces V* is still on the seabed, barely capable of movement and with a fading battery.

Yet as much as Messervy fears for the fate of Chapman and Mallinson, Carter senses that the commander has a secondary concern, one not too far behind the primary mission to save lives, which is that he and Vickers believe they have to be the one to rescue their own men. If the Canadians from Hyco or the Americans with CURV-III 'save the day' then, beyond the public delight and forced grins, it would be, to both Messervy and the company, a dark day of disgrace. 'He didn't want CURV to do the rescue, so he was virtually pleading to make sure that Oceanics pulled off the rescue.'

Carter listens as Messervy states, part order, part plea, 'You've got to get them.'

On the deck of the *John Cabot*, technicians have removed the protective metal panel to work on CURV-III's main connector system ahead of the remote-controlled vehicle's imminent launch. They used the sailing time out from Cork to the accident site to hook up all the interconnecting cables required to power and control the vehicle, but the work had been completed out on deck in foul weather, with waves slapping hard against the ship's hull and rolling on up the side, soaking the men. What they haven't appreciated until now is how much water vapour has managed to penetrate into the main connector system.

So, shortly before midnight, when Larry Brady and the team complete their work and power up, they watch in shock as electricity hits the water droplets and the system 'literally blew up ... with a lightning crack of arching electricity and roiling smoke, the 55-pin connector became nothing more than a blackened, smouldering mess.' On the verge of launch, CURV-III's entrance is – in a split-second – cancelled. As they begin to cool the system down and figure out the best and

The American team behind CURV-III. From left: John De Friest, Tom Wojewski, Bob Watts, Larry Brady, Denny Holstein, William Patterson, William Sanderson.

fastest repair, they know it isn't going to be a matter of minutes. It's going to be measured in hours.

Pisces II is on *Voyager* with an as yet undiagnosed fault, *Pisces V* is on the bottom with limited battery power, and now CURV-III is rendered inoperable just as it's been called into action. When the news is broken to Messervy, he's stunned and can't quite believe it. They now have less than 12 hours – maybe a little more if Chapman and Mallinson have managed to reduce their oxygen intake to a level even lower than before, but nothing that can be relied on. All Messervy wants to know is, how long? The answer that comes back isn't good. The Americans know there's no time to wire in a spare, so they begin to cut back the cables, run out the wires to identify their constituent jobs, put on solderless lugs and begin to hard-wire each of the 55 leads into the van's connector strip. They're working as fast as they can, but it's still going to take three to four hours – one-third of the submarine's remaining oxygen window.

CURV-III has been considered a back-up by everyone except Al Trice, who has no wish to rely on something in which he's got so little faith. He's always had a good working relationship with the US Navy, he just doesn't believe in the efficacy of their latest technology. Maybe it all stems from their inability to sink his prototype back in 1966, but from the beginning of this rescue mission – what Trice has taken to calling the '84-hour day' – he genuinely doesn't expect CURV-III to work. It has far too many components – and the more parts there are, the more opportunity for faults. Still, when he hears the news of the blow-up, the Canadian doesn't know whether to laugh or cry. It does, however, force him and Messervy to think about what other options are available, for at this precise moment they have nothing operable.

At some point over the next couple of hours a plan is conceived that indicates the team's level of desperation. If they're failing to secure a single line, what could they achieve with a net? A vague scheme begins to take shape that if all three submersible vessels remain inoperable by 10 am, then the team should utilise the nearby fishing vessels in a last-ditch attempt to snare *Pisces III* as they would a shoal of cod, and attempt to drag her to the surface.

Back on the *John Cabot* the Americans discover that among the casualties of the electrical conflagration is the compass. Larry Brady knows it's not absolutely necessary for finding *Pisces III*, as CURV-III has sonar and there will soon be a new pinger in place on the sunken sub, but he knows he'll need a directional guide to help him keep track of the turns in the cable, because if the rescue cable he's carrying down becomes entwined with CURV-III's own umbilical, then the vehicle will become stuck and unable to reach its target. So Brady hauls over his canvas dive bag, then fishes out his own diver's compass. If he can fix this to the starboard side of CURV-III's frame, the compass will be in clear view of one of the TV cameras. Once completed, he decides to leave the rest of the team to their work and head down to his cabin to try to get an hour or so of sleep. He'll be on stage soon and wants to be at his best.

* * *

Midnight is a lonely hour for the men in *Pisces III*, and
shortly after the witching hour their world becomes lonelier
still. *Pisces V* receives the order to return to the surface. For
the last three hours, since attempts to connect the choker to
the lift point were abandoned, Macdonald and McBeth have
kept the craft resting on the seabed a few feet from *Pisces III*.
During this time there have been a few brief conversations
over the underwater telephone, and the appallingly violent
noise as they tested their pinger, but for most of the time there
has been silence. Silent company is still company, and it has
been a comfort, one Chapman and Mallinson will only fully
appreciate upon its departure.

The reason for the retreat is CURV-III's delay. Messervy
worries that the estimated time of repairs may be longer than
four hours and wants *Pisces V* to return to the surface,
recharge her batteries and be ready to return by dawn. If
CURV-III is *Pisces II*'s back-up, then *Pisces V* will be CURV-
III's back-up. There can be no 'last line of defence'. They must
always have alternative options.

For the last couple of hours Macdonald and McBeth have
taken it in turns to curl up in a sleeping bag. Now that the
order to surface has come through, McBeth is hit by a feeling
of 'absolute dejection'. In his heart he feels they may have
been the men's only hope and that they have failed them. He
knows that on the surface the rescue is in disarray. Macdonald
meanwhile tries to reassure Chapman and Mallinson that
CURV-III is coming, but Chapman doesn't understand. Before
Pisces V lifts off the bottom, Messervy comes through on the
underwater telephone to *Pisces III*. *Voyager* is still rolling
through stormy weather and outside the bridge from where

he speaks all is darkness. But he's intent on bringing the men a little light. Seeking to spread encouragement, he assures them that although *Pisces V* is retreating, fresh reinforcements will soon be on their way. He tells them about CURV-III, about the wonders of American automation and that this submarine is manipulated by remote control from the surface.

But what he says has the opposite effect. From the few words that he can make out, Chapman is filled not with comfort but deep apprehension. Neither man in the stricken sub has any idea what CURV-III is, or what it can and cannot do, and so Chapman begins to worry that it will achieve nothing. If experienced pilots, veterans of an almost identical rescue to their current attempt, have been unable to fix a lift line after 18 hours on the bottom, what are the odds of a machine, even an American machine, being able to do so?

The departure of *Pisces V* is but a brief distraction from this world of worry.

The sand on the seabed begins to stir around the sub's skids as *Pisces V* gently rises up alongside *Pisces III*. If Mallinson and Chapman wish to they can stand up and look out of the porthole as the submarine's ghostly form drifts past. Instead they sit and listen as *Pisces V* says goodbye.

Thank you, says Chapman, then, concerned that the tone of *Pisces V*'s farewell was too final, adds more in hope than expectation, 'See you down here again soon.'

Macdonald replies that he doesn't think so, that most likely CURV-III will soon have them on the surface. Chapman tries to cling to the confidence he thinks he can detect in the Canadian's voice.

Pisces V continues to rise up, past the top of *Pisces III*'s bow. The further *Pisces V* climbs up into the dark, the smaller *Pisces III* – fixed on the seabed – appears, until the receding waters envelop her completely. McBeth sits silently at the controls. The 30-minute rise to the surface will be the worst journey of his life. He cannot shake the feeling that they are abandoning two men to their deaths.

Like everyone else on board *Voyager*, Harry Dempster is hoping for the best, but unlike anyone else he is also working out the best position to photograph both men if, as seems likely given the delays, they are recovered dead. He is concerned by talk about a plan to raise *Pisces III* from the *John Cabot*. For if, as suspected, he snuck on board one ship, he certainly won't be invited onto a second. Instead he will have to wait until the bodies are brought back onto *Voyager*, and then the most likely image will be of bodies on stretchers, covered with one of the ship's grey woollen blankets. He knows the key image will be either Chapman or Mallinson carried out of the submarine, but he's not sure if he's going to be able to get it. Making these plans isn't a betrayal of the two men and the entire rescue effort. It's simply his job as a professional newspaper photographer.

After being deposited on deck by the Sea King helicopter, an experience he recalled as 'people frantically grabbing at my feet', Geoff Hall, a bright 27-year-old with a taste for adventure and a facility for electrical repairs, makes his way down to *Voyager*'s recovery deck. He's a welcome sight, as the team could do with a fresh pair of hands. Hall looks around at the creased, lined faces of the crew and recognises that they are clearly exhausted. There's not complete despair, but there's despondency, mixed with desperation, and 'everyone was really down.' When *Pisces V* is brought back on board, the maintenance team discover that despite being plugged back into the power supply, the batteries fail to recharge.

Hall takes up position by *Pisces V*'s power panel and begins to think through the reasons for the problem with the batteries. An examination of them leads him to conclude that they appear to be in good working order and that the fault lies instead with the battery chargers. Hall is familiar with the design, as he himself had them installed. He goes down to the stern of the ship, to the compartment where the battery chargers are stored. The first thing to check is whether there's an adequate electrical supply. Hall knows that *Voyager* is predominantly a DC ship – running on direct current. She has electrical propulsion and the main distribution system is 220 volts DC. He also knows that, at some point in the past, probably during a refit at the dry dock in Manchester, the company had fitted some AC (alternating current) generators, which had fairly limited capacity.

The main battery chargers are three-phase – 440 or 450 volts – and when he checks, he notices that the main battery is 110-volt DC. He sees that the way the chargers are arranged – the ones that have been supplied by Oldham – means that they are using two chargers. Oldham don't make a 110-volt DC charger and it's common to have two chargers linked in series to provide the required output.

When Hall checks the power supply he discovers a problem: one of the main three-phase circuit breakers – the centre yellow face – has an open circuit, so the battery charger is only being fed by two phases instead of three. Hall asks around, and, despite a quick search, no spare can be found. He decides to improvise. As a temporary measure to get *Pisces* charged up, Hall isolates the power supply from the ship's switchboard. Although Hall knows there's back-up protection for the switchboard and the move should have no detrimental effect on the ship's communications, there's always the possibility of a glitch. He also knows there's no other option on the table. He puts a temporary choker across the defective phase on the circuit breaker and then restarts the power. He holds his breath.

It works. The submarine's batteries begin to recharge.

After his conversation with Messervy, Carter heads down to the aft deck where the red hulk of *Pisces II* is positioned after her emergency recovery. He begins to work through all the usual pre-dive checks and tells Geoff Woods to take a look at the aft sphere, from where the water alarm was triggered. Carter has an inkling that this could be the return of an old problem. For next to the aft sphere is the oil reservoir, which is accessible via a hatch on the floor. In order to check out or clear out the aft sphere the oil reservoir has first to be drained, and the maintenance team would loosen the screws and take the hatch off. It's not uncommon for one of the screws to fall into the bottom of the sphere where the water alarm – which consists of a pair of wires – is positioned. If anything touched those wires, water or metal, the alarm would be set off.

As Carter works through the long checklist – topping up the battery oil, ensuring the emergency buoy release is working, clock wound and set, drop-weight wrench in position, sonar system operational, ports cleaned, cabin vent shut, TV system functioning, and on and on – Woods is at work on the aft sphere. Carter is close to completion when he hears Woods holler. He climbs out of the sub to see Woods holding up a handful of screws and shouting, 'The water alarm!' They have their explanation and their solution.

Carter considers *Pisces II* good to go. Messervy is unsure and insists on a double-check. He can't have another cancelled dive or, worse still, more men trapped in a second sunken sub. Carter convinces Messervy that *Pisces II* is ready and is now their best shot at getting a working lift line on *Pisces III*. The next question is, who should take the dive? A discussion takes place on the bridge. Carter insists, as a fresh pair of hands and the man who is signing off on the sub's safety, that he will fill one seat. Among the volunteers, and the one with the most authority, is Bob Eastaugh, the senior operations manager. But then the only man who can refuse him, does so.

You're needed here on board, instructs Messervy. He then seeks to minimise Eastaugh's evident disappointment by lavishing praise on his expertise and skills that will be required during the upcoming lift. Eastaugh isn't buying it.

Instead Des D'Arcy is once again picked to pilot *Pisces II*. For him, this dive is personal. He has twice before attempted to rescue his colleagues, and twice failed, beaten back by fate. He can't let it happen again. It's settled that D'Arcy will act as pilot, while Carter will control the manipulator arm and position the toggle.

The decision made, Carter and D'Arcy return to the aft deck. Darkness has enveloped the vessel and the work lights in operation are dazzling. Carter is working out the best way to fix the Vickers rescue toggle on to *Pisces II*. The rope they had previously been working with has come from the *John Cabot* – it's stretchable and almost two inches in diameter. During yesterday's first dive the buoyancy of the rope was so great as to tear the toggle from the manipulator arm, so Carter is keen to try a new approach. While others around him suggest plastic tie-wraps, he dismisses the idea, worrying that once the toggle is firmly fixed in position into the aft sphere of *Pisces III*, *Pisces II* won't then be able to break away and leave the lift line. It had to be something lighter. Carter opts for simple white string, and uses it to bind the rescue toggle and its connecting rope to the manipulator arm and the torpedo claw.

D'Arcy, meanwhile, is stripping out all superfluous weight from the sub, such as the pan and tilt camera, and making sure they have extra oxygen cylinders on board in case they have to 'sit to one side' for an extended period. It's never explicitly mentioned, but perhaps in the back of both men's mind is the grim scenario that if Chapman and Mallinson cannot be rescued in time, they will not die alone.

At half past midnight in the situation room at Vickers in Barrow, a message is received from HMS *Hecate*, based on the information provided by *Voyager*:

> PII ready to dive in two hours, weather permitting. JOHN CABOT preparing CURV for launching – expected ready for launching in 3 to 4 hours. PV on surface being recovered to recharge batteries. Not available for at least 6 hours.

For Redshaw and Mott the disappointing news is digested with the last of the cold coffee and a sandwich platter now curled with age.

Almost three hundred miles away in London's Fleet Street, home to the nation's newspapers, the early editions are rolling off the presses. The *Daily Express* front page declares: '*Pisces*: One Last Chance', with a sub-heading: 'Vickers Chief: "It's neck and neck ... time is against us"'.

In Broom Cottage June Chapman has retired to their upstairs bedroom but sleep will not come. She spends hours wrestling in the sheets before deciding to get up again. She will sit through the night on the sofa, looking at the unfinished walls and wondering whether this will be the day her husband dies.

On the *John Cabot* one of the Americans, Tom Wojewski, is working under a heavy-duty arc light so hot that he's stripped down to his striped short-sleeve shirt. Right beside him is John De Friest. He's made of sterner stuff and refuses to get out of his boilersuit or even take off his dark-blue baseball cap. Together with Bob Watts, they're all wrestling with the spaghetti mess of cables and the central control panel of CURV-III. They're being asked every 30 minutes how long the repairs are going to take, and every time the answer is the same. They said 'three to four hours' at the start, and it's not going to take any less, interruptions or not.

During the two-hour period between 1 and 3 am, Messervy asks 'more than once for a "go" time for CURV'. The reply comes from the *John Cabot* that the team are 'not able to shorten the estimated repair time'.

At 3 am Messervy orders the launch of *Pisces II*. In no other circumstances would he do so. The weather, recorded in the log entry on *John Cabot*, is: 'Strong west-southwesterly breeze, rough sea and heavy swell, overcast, intermittent rain showers.'

First, in order to guide *Pisces II* to her sister vessel, a Gemini bearing a pinger bumps across the breaking waves to the bright orange buoy line connected by *Pisces V* to mark *Pisces III*'s exact location on the seabed below. The diver reaches over to the line and connects the pinger to the rope, then lets it drop down the line and through the fathoms.

Chapman records a surprisingly optimistic entry: 'Advised PII coming down at 0300. Setting timers at 30 min. for sleep. Very tired now but morale good (weekend!).'

A later note at 1.13 am reads: 'Heard that a pinger was being sent down.'

At 4 am *Pisces II* is swung over the side of *Voyager*, towed into position and eight minutes later begins to descend. Although the pinger has a head start and a direct route straight down the surface buoy's marker line, D'Arcy and Carter are unsure if it has already touched down on *Pisces III*.

It takes twenty minutes for *Pisces II* to sink through the shades of blue merging into black and alight on the sandy seabed, where the sub's sonar picks up a promising ping. D'Arcy directs *Pisces II* in the direction of the sound, and Carter peers through the observation porthole at the sand and the curtain of passing crud. They realise the sonar reading is indicating a mass smaller than *Pisces III*, so D'Arcy gently spins the sub around a complete 360° circle, her lights illuminating the waters, although passing fish show a complete lack of interest. Before the rotation is complete, the sonar picks up a blast big enough to be *Pisces III*. D'Arcy fixes the sub on the location of the sonar hit and, at a steady six miles per hour, *Pisces II* makes her approach.

At 4.30 am Chapman notices the water outside the porthole begin to lighten. He peers through and recognises the spotlights of *Pisces II* moving out of the darkness. In the notebook he records:

> PII obviously coming down very close to us and great excitement. We can see her lights. Tried not (underlined) to get to worked up about this as still lot to be done.

Thirteen minutes later he writes: 'Interrogated by PII's sonar. She must have been very close.'

Pisces II crests a rocky outcrop on the seabed and the spots light up their target. Carter said, 'We saw this white submersible sitting on its end. We approached it straight on and didn't have to negotiate round it.' Any brief feeling of relief at having located *Pisces III* is tempered by concerns that a careless approach will see their sub and her floating polypropylene line become entangled in the buoy line that now runs from the propeller guard grille to the surface. Carter tells D'Arcy to go easy, but D'Arcy has already seen the floating line and steers to avoid it by approaching the sub from the right-hand side at an angle of 70°.

Looking out of the porthole, Carter is checking for the aft hatch into which he knows he needs to drop the rescue toggle then back off, using the combined force and tension of the reversing sub to snap the string ties that bind the toggle to the manipulator arm, allowing both toggle and rope to remain wedged in the aft sphere ready for the lift. It takes about six minutes from when *Pisces II* discovers *Pisces III* to the point at which Carter is ready to make the initial approach. Cautiously *Pisces II* closes in on her sister sub with the manipulator arm and tethered toggle outstretched. The toggle sits across the manipulator arm like the crossbar on the letter 'T', and Carter plans to get close enough to rotate the arm 180°, forcing the rescue toggle to slide down into the open cavity of the aft sphere, which is roughly one foot square. He isn't displaying the nerves that anyone else would feel in such a position. As the single best chance of saving the lives of two colleagues, the pressure internally should match that outside the portholes, but Carter has a gift for shutting down extraneous distractions, silencing the chattering voices of doubt and doom. The task, to his particular mind, is as simple as putting a lock in a key and turning.

D'Arcy brings *Pisces II* as close to *Pisces III* as possible, a gentle combination of thrust on and off, with a careful eye on his greatest fear: a collision. By 4.46 *Pisces II* tells *Voyager* 'that it was manoeuvring around P.III'. The sub inches closer still, allowing Carter to position the toggle on the cusp of the aft sphere.

He touches the controls and watches as the toggle tumbles 'and just fell open'. The device – best described as like an upside-down umbrella – works perfectly, with its crossed arms now unfolding to lock under the roof of the sphere. The next cautious moment comes as D'Arcy backs away. The tension builds on the string attaching the toggle to the manipulator arm until it snaps and 'comes off easy'. Carter can see the toggle securely wedged in the aft sphere with the polypropylene line floating above the sub, rising up into the dark waters and on up to the surface.

It's 5.05 am and a rescue line is now securely fixed.

Carter talks to the 'two Rogers', estimating that he's only fifteen feet away from them but in an entirely different world, safe in an operational sub with a feasible means of ascent, while Mallinson and Chapman are breathing foul and dangerous air. It's immediately apparent that neither man is in good shape – according to Carter, 'They were barely responding or sounding well. They were not really awake, almost comatose. They sounded in really poor condition.' On the surface Terry Storey takes over the communications as they quickly exchange words for a handset click in response.

Pisces II's work is not yet done as Carter and D'Arcy move on to the next task. The previous day the choker had been sent down the first line *Pisces V* had connected to the propeller grille. *Pisces V*'s line has no lifting potential, acting only as a guide to the location of the sunken sub, but if the choker could be moved via the manipulator arm onto *Pisces III*'s main lifting point it would provide 'an uninterrupted heavy line to the surface'. *Pisces V* made repeated attempts to complete the tricky manoeuvre, but the sub's failing battery limited their ability. Can *Pisces II* now succeed where *Pisces V* failed?

Pisces II eases off and, like a dance partner, begins to spin around the sub, moving 240° to the propeller grille, where the choker is resting. The message sent to *Voyager* and recorded in the underwater communications log reads:

> P.II got soft noose in manipulator, attempting to locate lift
> hook – semi buoyant not much current but making things
> difficult – confirms that toggle firmly in after hatch. Will
> inform surface when soft noose in hook.

For the next hour up until just after 7 am, Carter pits his own
dexterity against the choker, which he tries repeatedly to position up onto the lift hook. He comes close a couple of times
but fails to make the crucial connection.

At 7.08 the message is relayed to *Voyager*: 'PII unable to
attach soft noose of choker onto main lifting point – advised
standby for permission to ascend.'

The message D'Arcy and Carter passed on, and which is
recorded in the log, reads:

> We have toggle in the aft sphere with a firm grip. Have had
> a go with a soft noose and pendant. Will be going up now
> to have a discussion with the surface.

At 7.20: 'PII off bottom'. The rescue sub tilts down as she
manoeuvres around her stricken sibling, as if extending a
respectful nod of the head, then initiates her ascent and begins
to rise. As the lights retreat, *Pisces III* is left alone once more
in the cold darkness. Inside, Mallinson and Chapman set the
timer for another 30 minutes – when they must rise to yet
again activate the scrubber – then settle down to another
brief sleep, one that this time is accompanied with the first
genuine surge of hope.

* * *

Messervy knows they have one line secure and in position, but he wants a second line in place before they begin, and *Pisces II* hasn't managed to attach the choker that would provide this. It's not just a matter of 'belt and braces': the first line has a breaking strength of fifteen tons and should hold, but it's not going to be an easy lift, not with these waves and *John Cabot*'s high bow line. If CURV-III can secure a second line, he will be a much happier man.

At 8.30 am Bob Eastaugh transfers over to the *John Cabot* to supervise the CURV-III operation, which they hope will be followed by the final lift. Messervy and Trice will follow shortly, but at present Eastaugh is enjoying a brief break from what has been a tense partnership. On arrival he immediately goes to meet the captain, and together they go to inspect the lifting equipment. Eastaugh knows the vessel has the capability to lift *Pisces III*, and since CURV-III is diving from the *John Cabot* it makes sense to lift from here. The issue is the positioning of the lift, which will be from the high bow of a vessel that rises and falls with the waves. After the check, he calls Messervy and lays out his plan. Messervy, to his surprise, is quick to agree and sets about organising the transfer of the lift line laid down by *Pisces III*, which must now be transferred from *Voyager* to the *John Cabot* by a Gemini RIB.

This has to be carefully done. If the lift line, for whatever reason, falls overboard as the Gemini is buffeted by the waves, all will be lost. So the top end of the cable is carefully unbolted from the winch mechanism, then transferred by hand to one of the divers, who sits tightly cradling the rope as the Gemini pilot slowly steers the rubber craft over to the

John Cabot. On arrival the loose end is carried through the ship, with the fixed end draped over the side and following behind, where it falls down into the water and 1,575 feet below remains fixed on *Pisces III*. The rope is then carefully manoeuvred over the bow sheaves and fixed to the giant winch line. Once in position, all involved can breathe a little easier.

Larry Brady is in a bunk on one of the lower decks and enjoying a deep sleep when one of the team arrives to rouse him. He dresses, stops at the canteen for a fresh cup of coffee, then heads back up onto the deck he feels he left only minutes ago. The story Brady is told is that the *Pisces* vehicles have been unable to complete the rescue and that, while *Pisces II* has managed to fit a rescue toggle into the aft sphere, the line is only sufficiently strong for '12,000 lbs', not enough to lift the sub and the flooded sphere. (In fact the line fitted by *Pisces II* can carry the sub, but because it's a risk, a second line is required.) As he remembers it: 'It was up to us to attach a line capable of lifting the lost sub safely.'

But the rescue toggle constructed by Vickers Oceanics, the first of which Ted Carter has successfully fitted into the aft sphere two hours ago, is now deemed too risky by Brady and Bob Watts. After a careful inspection, the Americans remain unconvinced the simple umbrella technique of flopping open inside the sphere will work: 'It was strong enough,' Brady recalls, 'but would it open, stay inside and hold?' This leads to an argument with Bob Eastaugh and the Vickers team, who resolutely stand by their design. Brady argues that what he wants to do is only going to increase their chance of success, so over the objections of Eastaugh he makes a last-minute alteration to the design.

He takes a large crescent wrench from the CURV-III team's tool box, then asks for a welding kit to be brought up to the deck. Pulling on a protective mask and firing up the torch, he begins to weld the wrench onto the centre of the toggle. He then attaches two strands of bungee cord that will act as springs, holding the toggle open, for he believes that without the springs the toggle could collapse and pull out of the sphere if the lift line ever goes slack. Brady wants to prove his new adapted design will work, so while he is welding, the team construct a replica of the aft hatch out of plywood. Bob Eastaugh watches in stony silence. In any other circumstance he would love to see Brady proved wrong, but there isn't the time and, as he reminds himself, they're all on the same team. The toggle slips inside and the bungee cords act as planned, holding the twin arms open. It works.

The launch of CURV-III is tricky. The handling equipment for the vehicle requires a 19-foot articulating crane capable of a 5,000 lb lift, as well as a hydraulic traction winch and a davit and block for handling the nylon line and main cable. Although the distance from the deck of the *John Cabot* to the surface of the sea is only 35 feet, to Brady it looks like a mile. He's used to working off a barge running low in the water, not a massive cable-layer such as this, and then there's the weather, which hasn't eased up. He's worried that the extreme length of the tag lines will not permit any stabilisation of the vehicle during launch, and that CURV-III will spin in the wind and rotate as the *John Cabot* rolls with the waves. It does just that, but although it swings and rocks, the lift lines hold and after a couple of minutes, at 9.40 am, it touches down into the Atlantic.

Brady is in control now, and everyone must follow his and CURV-III's lead. As he recalls, 'Whatever reason for the delay in launching CURV, politics or pride, the CURV was on its way. The Yanks were given a chance.' His kingdom is a 10-foot by 10-foot by 20-foot aluminium container, or 'van' as it's called, into which a SLAB-603 instrumentation system comprising all the electronics required to control the remote sub is fitted. Tucked in tight on his swivel chair in front of a bank of controls, he directs the sub towards the buoy with the *John Cabot*'s captain now steering the ship along in the remote vehicle's tiny wake.

On the *John Cabot* the bow sheaves of the cable layer continue to pay out the two-inch, double-braided lift line attached to the toggle, which is gripped in CURV-III's mechanical arm. In the control trailer Brady has to contend with more than a few 'back-seat drivers', as Bob Eastaugh cranes his neck to catch a glimpse of the two small TV screens relaying images from CURV-III's cameras. For the first few minutes there's little to see but rising crud and krill and the distant fin of a darting fish. Yet the mission will begin to move forward like clockwork. Brady is manipulating three 10-horsepower propulsion systems that enable CURV-III to manoeuvre underwater. One is mounted on the centre of the vehicle and used to provide vertical control. At a flick of a switch CURV-III begins to descend at 30 feet per minute.

9.50. CURV-III at 190 feet.

10.12. CURV-III at 1,100 feet.

As Brady begins to approach the bottom he points the sonar down, lines up the diver's compass strapped onto the starboard side of CURV-III's frame and begins to pass over what he calls 'the night-time desert' en route to the accident site. CURV-III floats with around 10 to 25 lbs of positive buoyancy, which is achieved courtesy of the blocks of syntactic foam fitted around the top of the rectangular aluminium frame. Once on the bottom, Brady uses the vertical thruster to drive the vehicle down and so counteract the foam floats. This allows the wash to move up and away from the vehicle, which gives the cameras a clear and unobstructed view. The sub also has a 250-watt mercury vapour spotlight that can both pan and tilt and bring a brief blast of daylight to this nocturnal world.

Brady has only been scanning the surface for a minute when the sonar picks up a sizeable hit approximately 250 yards away and, a minute later at 10.31, his cameras pick up *Pisces III* 'suspended like a child's top as it leaned slightly to one side, the sail above the pointed tail section with the open hatch and the flooded aft sphere resting on the bottom'. For Bob Easthaugh and the Vickers team it's the first time they've seen live real-time footage of *Pisces III*, and the moment is emotional, with Easthaugh smiling broadly but looking away to bite his lip lest it tremble.

Inside *Pisces III* Chapman and Mallinson are mentally addled, just emerging from sleep. Outside the sub they can hear the movement of CURV-III's motors, but they're confused as they believe the remote-controlled sub has already been and gone. As Chapman wrote in his notebook:

In touch again with surface. All well. Informed about CURV ... thought that CURV had already been down as we had heard movement around us.

Brady describes CURV-III as like the 'dog on a leash', and just now the miniature sub is 'sniffing' around *Pisces III*, moving up and down then swivelling around, in search of the best angle of approach. On the surface Brady is flexing his hands before re-gripping the controls, aware not only of the extra bodies crammed into the control container but those outside on deck, who can't quite decide whether to focus their attention on the rough seas under which the action is unfolding or the container in which Mallinson and Chapman's fate will be decided.

Brady is relaxing into his role. He's done this manoeuvre hundreds of times before, and he reminds himself that he can do it in his sleep, although he shakes this thought away lest it remind him of the exhaustion he's already enduring. He manoeuvres CURV-III around, then lines up the toggle on the arm with the open aft sphere. Brady can see the lift line and toggle inserted by *Pisces II*, which rises out of the sphere like a straw in a snow cone, but he knows there's ample room for this second line. He manipulates the controls, watches on the monitor as the arm moves out and slowly drops the toggle into the aft sphere. The springs then activate, wedging the toggle in place.

After the problems and mishaps that have dogged the rescue at every step of the way, no one in the control container can quite believe how smoothly the CURV-III mission is completed. As Brady writes:

It almost felt anticlimactic after all the travel, loss of sleep and anxious moments, but the insertion was really no harder than parking the car in the garage after a long drive home.

Brady pulls CURV-III back, the masking tape securing the toggle lift line to the remote sub, snaps off and the lift line holds secure. In a dig at Messervy and the Vickers team, Brady wants to be the first to issue the only command that counts:

'Bring her up.'

CHAPTER THIRTEEN

The tick-tick-tick of the egg timer started the breathlessness. The idea that each second past was a breath gone and one less breath to take. When would they reach their final breath? Time would tell, and time was not on their side. Time was the enemy. There would come a point, calculated in hours, minutes and seconds, in oxygen inhaled and carbon dioxide exhaled, when it would be time to die, and neither man was ready.

Inside *Pisces III* Mallinson and Chapman prepare for the lift. They call the surface to get the current atmospheric pressure, which has changed from the early hours of Wednesday morning when they first climbed inside. Chapman listens and resets the barometer, turns on the oxygen valve and hears a thin, weak whistle of air emerge and slowly push up the barometer needle. The last instruction from Messervy is to unpin the hatch before they leave the bottom. They understand that this is in case, by the time they reach the surface, they are already unconscious from lack of oxygen or knocked out from the trauma about to ensue. Or dead, thinks Mallinson.

It's 10.50 am. Mallinson and Chapman hear, 'You're being lifted now PIII. Over.'

Both men are in considerable discomfort, at times tipping into abject pain. The atmosphere is so thin and clogged with carbon dioxide that their headaches are severe and persistent, their thoughts a muddy soup of images and old emotions, and distractions from the surface are treated as irritating inconveniences. They switch on the internal light and Chapman's first thought is, curiously, how annoyed Bob Eastaugh will be at the mess they have made. The light illuminates a clutter of damage and debris marinated in the stink of shit and urine. Chapman begins, half-heartedly, to attempt to tidy up, and Mallinson lends a hand before both then put on their orange lifejackets and retreat to opposite sides of the sphere, jamming themselves against the wall in preparation for what they know is going to be a very bumpy ride. They look at each other, and then smile. The increase of oxygen added to calibrate the atmosphere to match the surface has helped clear their fatigued minds a little. It's easier to think, but the thoughts that come are not good ones.

Chapman broods on the expected angle of lift. If the *John Cabot* is not exactly above their current position then instead of a clean vertical lift, *Pisces III* will be dragged along the seabed. 'What if we are dragged and then the sub flipped over?' thinks Chapman. As he later recalls, 'Fear and excitement took charge and we began to burn up the oxygen.'

Mallinson is also worrying. He understands the structure of the sub like no one else, he knows her strengths and weaknesses, and the fact that the lift line is not where he would have placed it. The submarine's crash dive onto the bottom is bound to have weakened her structural integrity. The 'umbrella' – he still has no clear idea of what the rescuers on the surface are talking about and when he closes his eyes sees only an open-ended hook, like an umbrella's handle, from which *Pisces III* could yet wriggle free – has been slotted into the aft sphere, and he's fully aware that the only thing connecting the aft sphere to the front sphere where he now sits is a steel framework encased in fibreglass. In the dark he imagines the hidden cracks riddling the steel, and when the lift begins the frame cracking and *Pisces III* snapping in two, the aft sphere ever ascending while the front sphere tumbles back towards the bottom with Chapman and him trapped inside. He calculates the odds, and reckons the front sphere may yet survive a second uncontrolled accelerated descent intact. But what about the battery? Could it survive too? Their air would be out, the scrubber scrubbed out. His breath begins to quicken, speeding up the depletion of their oxygen supply as adrenaline floods through his system. Mallinson is no longer worried. Mallinson is terrified.

On the bridge of the *John Cabot* are Messervy and Al Trice, who caught a Gemini across from *Voyager* to supervise the lift. They are now looking out to the front of the vessel, where the great lifting gears are beginning to turn slowly. One full rotation is the equivalent of 20 feet of lift. Messervy instructs the mechanic at the lift controls to start slow, real slow. He tells him inches, not feet. Slowly the wheel begins to turn, the two lines go taut, tension ripples down 1,575 feet to the aft sphere of *Pisces III*, where the steel toggles begin to slowly tilt up. A passing flounder providing a fish-eye view would see the toggles move from a 45° angle to 75° then to 90° as the strain builds up and the resistance of over two tons of submarine bears down on a crucial piece of kit that less than three days ago was only a pencil line technical drawing at Vickers Oceanics. On the seabed the sand in which the white nose of the sub is buried begins to shimmer and tremble as a dusty blizzard of grains is shaken up and begins to float off with the current.

As the sub is being lifted by the tail, the first sensation felt by the men is a mild creaking and forward motion of the front of *Pisces III* tilting forward. Chapman looks at the depth gauge. It does not move. Yet there's forward motion. As the angle of the sub hits 50° to the vertical, *Pisces III* begins to move on the horizontal axis. She's being dragged along the seabed, just as Chapman feared. He believes the *John Cabot* is wrongly positioned and reaches for the underwater telephone and begins to shout with rising desperation: '*Voyager* ... *Voyager* ... This is PIII, over.'

There's no response.

He tries again.

Again, no response, only a whirring sound over the loudspeaker. They are now at 40°, then 30°, and just as Chapman is convinced *Pisces* is going to slam into an unseen rock or flip over, the depth gauge begins to flicker slightly. They're no longer only moving across the seabed. *Pisces III* is rising.

Relief is brief. *Pisces III* is off the seabed and at the mercy of motion, both the umbilical yank of the *John Cabot* and a pendulum swing that begins gently, a lullaby rock of back and forth, but then begins to increase in strength and momentum. Mallinson pushes against the wall and begins again to calculate the odds of structural damage. It's then that the first flip occurs.

On the surface the *John Cabot* is driven up by the black water of 20-foot waves, then plummets down into the wave trough. The cable wheel rises up, pulling *Pisces III* like a yanked fish, before dropping her down again a few seconds later. As they have nothing on which to grip, Chapman and Mallinson are hurled against the bulkhead and clatter onto the floor of the sub in what Chapman later describes as 'crazy upside-down world of noise, foul smell and fear'.

The depth gauge reads: 1,400 feet. *Pisces III* has risen 175 feet. Chapman is thinking about the 'hook' – as he visualises Vickers' toggle – and wonders how soon the motion will shake it loose. We will fall, he thinks. We will fall. Then the sub is struck by another violent upward swing that throws him to the opposite bulkhead, before he slips and falls backwards. Every movement sends a spike of anxiety coursing through his body, accompanied by the coppery taste of fear, like a mouthful of pennies. The motion is relentless, and the next 400 feet of vertical ascent an endurance test Mallinson and Chapman fear they will fail.

The depth gauge reads 1,000 feet. Chapman is shouting for the lift to stop. Both men know that if the toggle fails a drop of 400 feet could bust *Pisces III* wide open. They wouldn't die of suffocation but drowning. They can see the sub's skin splitting and the onrush of icy salt water swiftly filling the sphere, rising from ankle to knee to hip to chest to neck, their last desperate breaths taken pressed against the ceiling until the water rises above their heads, their lungs can no longer hold and any final breath gives way to a reflex swallow of the sea.

These thoughts of a not-too-distant death are distracted by a more immediate concern as the sound of water fills the sphere. Chapman spins round searching for the source of the leak, scanning the bulkhead walls for an ingress of water when his eyes spy the plastic shopping bags that have been their portable lavatory. Both bags, previously quite full, are punctured and spraying an arc of stale urine over the sphere. There's nothing they can do. It's just more mess and filth to disgust Bob Eastaugh.

In the dark-blue waters of the Atlantic *Pisces III* is rising slowly and with company. CURV-III is attempting to rise at the same rate but the remote-controlled vehicle's umbilical is beginning to coil around the central lifting line for *Pisces III*. In the brown metal shipping container on the foredeck of the *John Cabot*, Larry Brady tweaks the joystick and thruster controls and attempts to keep CURV-III always just a little ahead of *Pisces III*, a companion keeping a close eye, not a liability.

He is unaware that the cable is slowly tangling.

It's 11.37 am. The lift has been running for 43 minutes. The depth gauge reads 650 feet, over halfway to the surface, but the shorter the lift line becomes, the more traumatic and violent the turbulence as the elasticity in the rope reduces. Again Chapman, fresh from being smashed against the wall, reaches for the underwater telephone and screams for the lift to stop. There's no response. And now both men hear a familiar and troubling sound, not from the underwater telephone but from inside the sub. It's the sound of an oxygen leak. Mallinson figures that during the sub's latest contortions at the end of the lift line a piece of kit has, somehow, slammed into the valve handle of the main air supply to the pilot's console. Mallinson is positioned nearest the valve, and he manages to reach over and turn it off. The hissing stops. The questions both are thinking to themselves are how long was the valve open, and how much air have they lost?

In the small cabin of *Voyager* that serves as the communication room, Terry Storey, who is monitoring the radio, hears a voice through the crackle. It begins faint but becomes clearer over the next few seconds. 'It's Chapman and he's screaming. Chapman is shouting to "Stop the lift". He says they're going to die before they reach the surface, that the thrashing inside *Pisces III* is too much.' Storey immediately passes on the message to the *John Cabot*.

The time is 11.42 am, the depth 350 feet. The narrative here is confused. Chapman in his account recalled that communication was 'non-existent' but for 'garbled messages'. They pass on their current depth and wait, trying every minute or so to re-establish contact with the surface. First Chapman mans the telephone, then Mallinson. The pitch and tumble is distressing. The muscles in their arms and legs are burning from the strain of trying to lock themselves in position against the wall. The fear, expressed in tight shallow breaths and battery acid in the belly, is rising in direct proportion to the fall in oxygen. Over the telephone they hear a snatch of instruction: '... lifting ... shortly. Trouble with ... confirm that ... unpinned. Over.'

Chapman, according to his account, shouts back: 'Surface, for God's sake keep lifting ... crashing about. Hatch is unpinned. Please lift.'

In the shipping container on the *John Cabot*, Larry Brady, who's beginning to feel the telltale signs of an oncoming migraine, spots a problem. The umbilical cord that controls

CURV-III has now become badly entangled with the lifting line of *Pisces III*. This is another reason for stopping. To untangle CURV-III's cable from *Pisces III*'s main line will require no little amount of nautical dexterity. He needs CURV-III to retrace its steps and somehow unthread the needle, freeing itself from the line in which it has become trapped. He stretches out his already tense hands and settles back onto the controls. 'C'mon baby,' he mutters under his breath. The black-and-white footage on the 12-inch monitor shows the main lift line with CURV-III's cable looped around. Larry activates the thrusters and watches as CURV-III tries to move away from the cable, but instead it comes to a strained stop. Like a dog tied to a lamp post, it doesn't have far to go.

CURV-III is stuck, no longer capable of rising using its own power. On the surface Messervy wants to cut the power cable to CURV-III in an attempt to disentangle it, but when word reaches Bob Watts, who is on the bridge, he races down to remonstrate. He explains CURV-III may yet be needed if *Pisces III* breaks loose and falls. Instead it is agreed to find a way to lift both CURV-III and *Pisces III* simultaneously.

The short line has made the swings and near-summersaults of *Pisces III* so violent as to induce nausea. Chapman feels that he is going to throw up. His mouth becomes watery as the waves of sick rise up, then, at the last minute he swallows it all down.

The depth gauge flickers back to life. The dial slowly moves.

300 feet.

200 feet.

For the last few hundred feet the dark of the ocean has been diluted by sunlight. Yet it's not until *Pisces III* reaches

200 feet that Chapman realises this when he looks out of the porthole. He's immediately filled with hope: 'Daylight. Actual daylight began to stream in.' But it's said to be darkest before dawn, and the tribulations of *Pisces III* have not yet passed. There are more shadows on the immediate horizon. *Pisces III* rises to 60 feet, then immediately drops again to around 100 feet. Chapman can't help but think that they have been here before, hanging suspended 60 to 100 feet down beneath the mother ship; all he can pray for is that the cables, which have taken them this far, will hold on a little longer.

On the deck of the *John Cabot* Messervy isn't so sure. He knows they need to fix another line onto Pisces. While suspended under the water, *Pisces III* weighs two and a half tons but as she surfaces, her weight in the open air will increase to 12 tons and they need to get another lift line on to her, both to reduce the violent motion and so they can physically lift her out of the water. Now that CURV-III is out of action, there's only one other option. They need to send in the divers. A team of US Navy divers from the USS *Aeolus* have volunteered to assist as the dive team from *Voyager* have been working for almost 36 hours without a break. Yet an initial attempt by two American divers to dive down and fix a lift line is abandoned on account of the violent motion of *Pisces III*, so the divers from *Voyager* are called back in. (Mallinson remembers seeing one diver through the porthole with 'flaming blond hair' and thinking, for a second, 'Is that a mermaid?')

* * *

Bob Hanley feels the Gemini rise and fall with the waves, and looks up at a cliff face of steel – the stern of the *John Cabot* – and how it too is rising and falling at least 20 feet from the crest to the basin of every passing wave. He believes Chapman has called the lift off – for now – so frightened are both men of being dashed to death inside the sub. It's now his job to help stabilise the sub with a new line. The message to Hanley from Messervy over the handheld VHF radio is direct: Go in the water. Take it down. Connect it up. The message is simple, the task anything but.

He spits into his mask, wipes the saliva around the glass to coat it in a thin film, then bends to rinse the mask in the salty water before fixing it onto his face. He puts the regulator in his mouth and takes easy breaths, slow and steady, and finally rolls backwards off the Gemini through the crest of a rising wave and down into the blue. The first jolt of cold Atlantic water freezes him for a second, but he's done it enough times not to be fazed. The cold water flooding into his wetsuit is already beginning to be warmed by his body heat. The bubbles clear and Hanley looks down, kicks back his fins and begins to dive. The visibility is less than 60 feet, so all he can see is the blue-green waters and the shoal of plankton that accompanies each dive like dandruff or dust mites, and at the centre of his field of vision the cable, running down towards an inverted horizon, like a straw punched through a pale blue airmail letter. He kicks his fins, diving deeper down, and out of the blue begins to emerge a mass of white, soon the recognisable shape of a metal mass in violent motion.

It's *Pisces III*.

Hanley sees the cold white shape of the sub against the blue backdrop. Sunlight punches down past 160 feet, but as the sky above is all cloud and storm, the waters are darker, murkier than he hopes, but with his eyes fixed on the target he swims closer. He sees the first heavy lift line attached to the aft of *Pisces III* and for a few seconds the line is straight and taut and the sub hanging straight down at a 90° angle, like a pretty pendant on a chain.

On the surface, the crew up on the bridge of the *John Cabot* duck as the ship slams into another big wave, driving the vessel up 20 feet to the wave's crest. Floating 160 feet below, Hanley watches as the lift line immediately jerks up at least 20 feet and, like a yo-yo on a string, *Pisces III* sweeps back. If it were the hand on the face of a clock it would run backwards from 6 pm to 12 noon. Time, thinks Hanley. Everything is about time. After a second standing almost vertical, the momentum fades and *Pisces III* falls through an arc of over 180°.

Hanley knows he has to time his approach in synch with the waves to increase his chance of success. He needs to get to the steel fin by the sub's hatch where the steel lifting hoop is positioned. He kicks his fins and swims in, the heavy D-shackle cradled in his arms, hope in the shape of welded steel. On his approach *Pisces III* begins another jolting rise but Hanley is ahead of her, swimming not to where she is, but where she soon will be. He makes contact, smacking against the side, then struggles to get astride the fin. As the sub moves, Hanley gets his free hand onto the lifting gear and pulls one leg over till he's sitting on the sub when she begins to fall back. (He will later describe the position as 'a cowboy on a bucking bronco'.) *Pisces III* is reaching the apex of her journey, jolted by the *John Cabot* as she rises up the next wave, with Hanley tilting backwards as the front shoots up towards daylight. He tenses his thighs and digs his knees into both sides of the sail, which holds him in position as, finally, the pendulum swing turns him and the sub upside down. Hanley stares into the deep.

Fitting the D-shackle is difficult. A number of times Hanley jams his hand between the bracket of the lifting gears and breathes through the jolting pain, a violent stream of air bubbles bleeding from his regulator. The D-shackle comes in two parts, the steel backward 'C' and the crossbar 'I' that must be screwed in place to secure the two parts, which is, for the moment, attached to the 'C' by a short length of rope. Hanley finally feeds the 'C' through the lifting bar, then fumbles to find the crossbar screw as the sub rolls through another contortion. He fumbles again and, for a second, is distracted by the image of Chapman and Mallinson, less than one foot beneath his body, sealed in steel and fighting for fading breaths. The image strikes at the dry tinder of low-level anxiety that accompanies every stroke, but Hanley knows how to dampen it down and it doesn't catch. Instead he calmly focuses on catching the 12-inch bolt and threading it through the steel eye. This attempt fails as the sudden pivoting of *Pisces III* sees the bolt strike the eye's outer rim. Another attempt fails. A third. A fourth. Hanley loses count on which attempt the bolt slides through the first eye and into the second as he forces the screw tight.

All this time a second diver is holding back the coiled lengths of the new lift line rope, giving Hanley ample slack with which to work and preventing the line – and the D-shackle – from shooting out of his grasp. The D-shackle is now fixed. Hanley shakes it again to check, then kicks away from the sub and follows the line back up to the surface. He breaks through a crashing wave and pulls out his regulator. He shouts out, holding up his right hand with his thumb and forefinger in a circle and the three remaining fingers raised, the international symbol for 'OK'. The line is secure.

The pilot of the Gemini radios the bridge of the *John Cabot* to confirm the line is fixed. The message comes back: 'We're sending down another rope.' Hanley thinks Messervy is opting for a 'belt and braces' approach, and this is now one more belt for good measure. The rope feeds down from the *John Cabot* and passes to Hanley, who is joined in the water by Ken Brumby. Together the two divers swim back down to *Pisces III*. As the white sub's shape becomes visible, it's quickly clear to Hanley that the line he fitted has reduced the turbulence. 'Ken was brilliant,' he recalls. 'He was holding the line and we had to take slack out of the rope as the submarine was coming up and down and he got the pin, and then it was just a matter of bolting it up.'

The two divers swim back to the surface. Messervy now has his lines, and he instructs Mike Bond, another of Oceanics' divers, to swim back down to tell Chapman and Mallinson that the lift is about to recommence. Bond sinks below the surface and two minutes later taps the four-inch-thick glass of *Pisces III*'s observational portholes. Bond puts his thumbs up, then points to the surface and the sunlight. (For Hanley, the dive will not be without incident and some pain. Later, when the Gemini is lifted back onto the *John Cabot*, he fails to notice a swinging block and tackle, which smacks into his face, breaking his nose. Everyone will think the injury occurs while fixing the line on *Pisces III*. He will wish it had: 'It's a better story.')

The two Rogers' breathing is now completely out of control. Fear and, at times, genuine terror have increased their average breath count from 12 to 15 breaths per minute to over 25. At the beginning of the lift, almost two hours ago, their oxygen was at 800 psi, but it has fallen to 200 psi. Messervy and the rescue team calculate – wrongly – that the pair have managed to conserve enough air to last another five hours, to maybe 6 pm. Yet the men believe there's nothing like that much left. As Chapman later wrote:

> Two hours of the treatment we had suffered during the lift accounted for [a] 600 psi fall in oxygen; at that rate some 20 minutes more dangling around with the lines entangled would have left our ration of oxygen in the cylinder at about zero.

Messervy and Trice are on the foredeck of the *John Cabot*. The divers need more rope to help stabilise the sub, but by mistake the ship's crew lower down a large section of steel rope, which is too sharp and problematic for the divers to properly handle. Trice spots the problem and grabs a nylon rope, which is immediately lowered down. The diver quickly attaches it to the *Pisces III*'s lift point, but along the way the line has taken a right angle around a sharp edge, and as soon as the crew tighten the line the steel edge acts as guillotine, slicing through the line, which suddenly whips back on board. Trice yells to Messervy to duck, and he does so just as the rope swings past, reddening the side of his face.

Trice then shouts down to the divers to get ready. For all he knows, Chapman and Mallinson are going to be drowsy, oxygen-deprived and maybe even unconscious by the time the sub breaches the surface. The divers need to be ready to jump in there and haul them out.

As Mallinson recalls, 'When we were on the surface, it was just as dangerous as on the bottom.' They are again told the day's atmospheric pressure, and Chapman blows in a trickle of the remaining air and pushes at the hatch. It doesn't move. Mallinson pushes. It doesn't move. Mallinson figures that the crash landing on the seabed must have knocked the sphere's hatch out of its spherical shape, then 880 lbs of pressure per square inch of sea water has jammed it in place, and now, upon returning to normal atmospheric pressure, the hatch is refusing to yield. They put in a little more air, then a little more. Mallinson feels the panic rise again. They are so close, a hatch width from a safety that is still out of reach. He and Chapman are banging on the hatch. Then Mallinson tries another approach. He lies among the clutter and filth on the sub's floor, props himself up on his shoulder blades and begins to kick the hatch with his feet. He kicks once, twice, three times. On the surface Hanley is also straining at the handle, when suddenly Mallinson and Chapman hear a loud bang as the hatch opens.

Once again straddling the sail of *Pisces III* is Bob Hanley. He pulls the hatch back to 90° and clips on the safety line

Pisces III seconds after finally reaching the surface.

that secures it to the back of the sail. Packs of inflatable rubber bladders are passed over to him, and he begins to place them into the flooded aft sphere to drive out the water and stabilise the sub. The hatch now open, what Mallinson will most remember is the briny 'smell of seaweed'.

'Then we had our only argument.' The question of who should be first to exit is not the issue, it's who should be last. As the submarine's pilot, Mallinson feels the code of the sea and nautical ethics dictate that he should be the last one to leave *Pisces III*, but Chapman adopts a more pragmatic approach. The sea into which they have surfaced is wild, with heavy waves buffeting the sub. Chapman is concerned that if a crashing wave floods in through the hatch triggering *Pisces III* to sink, he's in a better position to make an emergency escape.

Chapman tells Mallinson: 'Roger, you can't swim.'

On the deck of *Voyager* the 'boilersuit brigade' gathers. The radio room has given a running commentary on every new development, and as the word filters around the ship that *Pisces III* is only feet from the surface, men emerge from bulkheads, hatches and doorways to line the deck and look out towards the expanse of the Atlantic, then down to a small square patch of sea by the bow of the *John Cabot*. Each man has been a human link in a chain forged under immense pressure over three days and nights, one that has pulled the submarine from the darkness up into the light. By their side is Harry Dempster, shooting off rolls of film on a camera equipped with a telephoto lens. When the white mass of *Pisces III* punches through the surface, a few spectators experience an immediate, uncontrollable emotion. Men whose eyes have been dry for decades begin to sob like children.

CHAPTER FOURTEEN

The dolphins have gone. Mallinson is hoping to see them on the surface, but he never does. The source of the squeaks and clicks that interfered with many of their communications have departed. In their wake, Mallinson and Chapman are greeted by a flotilla the likes of which they have never seen. First there's the red cliff-face of the *John Cabot*, towering above them, and then, a few hundred yards away, the familiar sight of *Voyager*. But it's the battleship grey presence of the British naval vessels *Hecate* and *Sir Tristram* and the USS *Aeolus*, as well as the miniature fleet of fishing vessels, that bring home the scale of the trouble they have triggered.

Chapman has no sooner looked around, his eyes struggling to adjust to the brightness of day, than a roar prompts him to lift his head up, as an RAF Nimrod flies overheard. The Nimrod, which has provided faster on-scene communication back to Britain, is not alone in the sky. A number of small planes, charted by the press, criss-cross the scene, carrying photographers hidden behind their telephoto lenses. Piloting the Gemini is Roy Browne, for whom Mallinson had once built a miniature steam engine during their downtime at sea, his delight at their survival obvious from a smile that won't go away.

Instead of being lifted by ropes straight up onto the *John Cabot*, it had been decided that the men should travel directly back to *Voyager*. It's an understandable decision, but one that will breed resentment among the American team, who will never meet the men whose lives they've helped to save. Browne steers the Gemini towards *Voyager* past a floating metal hulk. When Chapman asks what it is, he's told that it's CURV-III, still awaiting collection after its mechanical 'heroism'.

As they approach *Voyager*, Chapman begins to feel uneasy at the prospect of the Gemini being lifted up onto the ship's deck. Both he and Mallinson have had enough lifting for a lifetime, and he requests that they approach by the stern, where the divers hoist themselves back on board. It will require more effort, and Browne is concerned whether the men have the strength after their ordeal, but Chapman insists that they do.

The hanger deck and launch area are a sea of cables and ropes, spools of wire and oil-slicked shackles, oxygen canisters and diving regulators, buckets of masks and fins, tool boxes and battery packs. At the centre sit the familiar red and white shapes of *Pisces II* and *Pisces V*, along with their army of staff, all of whom want to shake the pair's hands or just stand in front of them and smile in a mixture of relief and wonder. For Mallinson, the clutter and disarray of the hanger deck resemble an operating table after a long, hard but successful battle to save a patient, the physical evidence of everything thrown into the fight. Yet it's a strange sensation, seeing the objects of one's salvation in an ordinary setting, not framed through a porthole window looking out onto a matter of life and death.

The press of people is too much for Chapman, who slips away to his cabin. On the wooden desk he finds his pilot's log, sitting where he had left it in the early hours of Wednesday morning. It's open at the page where the details of the dive are to be recorded. He sits at the desk, looks at the page and thinks for a second, quietly calculating how many hours he has been down. He writes, 'Pisces III Dive and Rescue', which he underlines. Beneath this he records:

Pilot Roger Mallinson,
obs [observer] Roger Chapman

He then writes: 'lowest dive, deepest rescue'. The pilot's log has columns headed 'Depth in feet', 'Duration in minutes' and 'Equipment'. Chapman writes in each respective column: 1675; 84½ hrs; PIII.

He notes to himself that this single dive is longer than the combined length of all his previous dives in a miniature submarine.

Yet there's no time to linger. Harry Dempster is anxious for some personal time with his new 'models'. The photographer had already walked into Mallinson's room to take his picture and, according to his subsequent newspaper report, Mallinson said, 'It's great to be back. Do you think you could get me a cup of tea?' Dempster says sure, then tells him that the entire world has been waiting to see them surface and that they've received telegrams from well-wishers from all over the world.

Divers on *Pisces III* after surfacing.

Mallinson smiles and says, 'So I gather.' Then he adds, 'But the greatest kick we had in the last few hours before we came up was to hear that the Queen had sent a good luck message on behalf of the Royal Family.' No one has yet broken the news that it's all a great misunderstanding.

In Dempster's account for the *Daily Express* he describes the scene:

Hollow-eyed and weary they lined the deck to welcome the two submariners aboard early on Saturday afternoon. It was fascinating to watch – the end of an incredible sea epic. What struck me as I gazed at the faces of the elated rescuers was that the two men who had spent so long entombed at the bottom of the Atlantic looked fitter, more rosy-cheeked, than any of us aboard. There were emotional scenes aboard. After I shook hands and congratulated Mallinson and Chapman, many hardened deck hands were close to tears.

Dempster is determined to get the best shots he can. He insists all the rescuers gather in one of the cabins and toast Mallinson and Chapman with cans of McEwan's lager. Mallinson holds a can but doesn't drink. Instead he waits till someone fetches him a can of lemonade.

Both men are examined by an elderly doctor who has flown in from Cork. When the doctor reaches out his hand to Chapman he tries to shake it, only to realise that the doctor isn't trying to congratulate him, only to check his pulse. Both men are judged to be remarkably well in spite of their ordeal, although Mallinson is suspected of still suffering from mild hypothermia. Otherwise the doctor's verdict is a single word: 'Incredible'.

After their check-up, Chapman goes in search of Captain Len Edwards and finds him in his small cabin, lying down with his socks and shoes off, his bare feet in the air. He has been standing at the bridge since the early hours of Wednesday morning, with only the shortest of breaks, and the pressure on his feet has caused them to swell considerably. Despite the pain, he cannot conceal his delight at seeing Chapman in the doorway, and shouts: 'You buggers!' He says that he knew where they were all along, and that too much time was spent trusting the electronic sonar and not enough listening to his intuition and old-fashioned seamanship.

Since almost the second they arrived back on board, the chef has been insisting that he will rustle up whatever either man wants to eat, they only have to say the word, but despite eating almost nothing for the past three days, neither Chapman nor Mallinson has the appetite for anything more substantial than a cup of tea or a sip of lemonade.

On the *John Cabot* Al Trice needs to sit down and rest. As soon as he sees the men are safe it's as if the plug has been pulled out of him. The relief is so intense as to be briefly disabling. Bob Eastaugh is supervising the recovery of *Pisces III*. The submarine can't be left where she is, dangling down from the ship, and will have to be towed back to *Voyager*. He uses the portable radio to tell Ralph Henderson that he needs more divers to come over and assist.

Messervy is anxious to get back to *Voyager*, but Trice says they should first thank the captain of the *John Cabot*: Captain

Back on board *Voyager*, Roger Mallinson and Roger Chapman celebrate with their rescuers.

Gordon H. Warren. When Messervy and Trice return to *Voyager*, Trice goes off to find Macdonald and McBeth while Messervy goes to see Mallinson and Chapman. He doesn't say anything to either man. He just stands and smiles, still wearing his lifejacket. He literally can't speak.

After less than two hours back on the ship, it's time for the rescued men to be returned home on the Sea King helicopter that's been dispatched to collect them. Mallinson and Chapman are joined on the flight by Messervy and Trice, as well as Harry Dempster. The photographer, like the rest of them, is winched up into the helicopter, but he holds his camera bag with its rolls of undeveloped film like his life depends on the bag, not the wire winch. Messervy is adamant that Bob Eastaugh join them on the first helicopter back, a mark of his grudging respect, but Eastaugh wants to stay and return with the subs. Messervy is insistent, so when the Gemini piloted by Ralph Henderson and carrying Eastaugh back to the ship runs out of petrol in the mid-Atlantic, the helicopter pilot moves into a hover and lowers the winch.

After picking up the final passenger, the Sea King arcs over the scene. The flotilla is already breaking up, the ships and boats moving on.

Henderson is left to wait, floating powerless in the sea, until help finally arrives.

At the first press conference back in Barrow, Redshaw and Mott announce to the assembled reporters that the rescue has succeeded, that the men 'look fit enough to play football' and that they 'send their love to their wives'. When asked about the crucial intervention by the American team, Redshaw replies: 'Our pride has certainly been touched a bit. Yet we have learned something from all this.' He goes on to describe the rescue as 'almost unbelievable'. Later, when Vickers are directly asked about claims that the rescue would have been 'impossible' without the use of CURV-III, a company spokesman says that this is a question they cannot 'specifically answer'.

Later Lord Robens, the chairman of Vickers, sends a message to the Oceanics staff:

> From all your colleagues on the board of Vickers Limited, I send to you and your devoted team at Oceanics our congratulations on the success of the tremendous efforts made to rescue Mr Mallinson and Mr Chapman. Throughout those long and anxious and difficult hours we saw a combination of total determination, steady nerve and inventive ingenuity. The outcome was a rescue which will always stand high in the annals of seabed operations, as will the cool, courageous fortitude of Roger Mallinson and Roger Chapman themselves. Will you please convey this message of congratulations and admiration to all who contributed to this memorable rescue, both at home and from across the Atlantic.

June has been asleep in the armchair when a call comes through to the cottage at around 5 am to say that the first lift line is in place. She goes upstairs to bed and quickly falls back to sleep. The next morning is visibly brighter, and before driving back into Barrow to Oceanics, she first stops at the post office in Dalton. Today is her parents' 30th wedding anniversary, as well as the wedding day of two close friends, a wedding they had both expected to attend, and June wishes to send telegrams of congratulations. At the post-office counter her pen hovers above the pad for an extra second before she steels herself and writes: 'June and Roger'. (The bride says later that she refused to leave her room and walk down the aisle until she heard the news that Roger was safe.)

After arriving at the Oceanics office, June listens to every update as the submarine inches towards the surface, but when *Pisces III* is suspended just below the surface while extra lines are fitted, she goes for lunch in a nearby pub. It's here, over beer and sandwiches, that the news is broken that both men are now, finally, out of the submarine.

The experience for Pamela Mallinson is more difficult. Already distrustful of the press on account of their behaviour while she was in her home, as well as the mistaken report about her husband's condition, she's unwilling to believe any news, even when it's good and comes from a trusted source like her chaperone Maurice Byham, who does eventually convince her that both men are now safe.

The Sea King helicopter touches down at Cork airport late on Saturday afternoon. In the film footage shot by the Associated Press we see Roger Chapman in white jumper and suit jacket coming first down the steps, followed by Peter Messervy in blue jacket, jumper and open-neck shirt. Mallinson steps off, smartly dressed in fresh white shirt and knotted tie. All are sporting the British 'stiff upper lip', and there's a determined effort to play down almost every aspect of the rescue,

Roger Mallinson, Roger Chapman and Peter Messervy (far right), celebrate with champagne for the press after their arrival in Cork.

including the question of the air supply, which has, quite suddenly, become more than adequate.

Walking into a scrum of reporters and cameramen, whom the uniformed Irish police attempt to hold back, Chapman is asked: 'Can you tell us how you feel?'

'Very well, thank you,' he replies.

When asked the same question Messervy just gives a clipped, 'Fine.'

After Mallinson climbs down from the helicopter, he's asked if he had any doubts if he'd ever be rescued. He smiles and deflects the question: 'We had a good team on top.'

'What was the worst part?'

'Coming up.'

Surprised, the reporter asks, 'Why would you say that?'

'Very rough.'

When Chapman, who's now standing beside him, is also asked if there was ever a time when he felt he would not be rescued, his answer is more expansive, but robust and professional: 'Not at all. We had about a day's supply left for life support, we knew what was going on, we could talk to the surface all the time, so there was no trouble at all.'

This is the first time that the question of their oxygen supply is broached, with the quantity revised from what was understood to be the case during the incident.

REPORTER: Did you have messages from your family?

CHAPMAN: Yes, and also from the Queen on the first day, which was very nice.

REPORTER: You got a message from the Queen?

CHAPMAN: She just wished us all the best …

SECOND REPORTER: Can you tell us what conditions were like down there?

CHAPMAN: They were very quiet.

MALLINSON: They were very pleasant.

SECOND REPORTER: Did you sleep?

CHAPMAN: No, we couldn't sleep much, but we nodded off.

SECOND REPORTER: Late last night, when you learned that they hadn't been able to fix the heavy lifting gear on, did you feel a crisis then?

CHAPMAN: No. No. Not at any time.

A shadow seems to pass over his face as he answers, and immediately he begins to move off.

Out of shot someone shouts: 'Give them a bit of space. Everyone back here, please.' The photographers line up as Chapman, with Mallinson beside him, is handed a bottle of champagne.

CHAPMAN: Shall we pop it now?

REPORTER: Commander Messervy, was it more rum than you thought it would be?

MESSERVY: Yes, it was.

REPORTER: In what way?

MESSERVY: The weather was bad.

He then takes a long swig from the champagne bottle.

REPORTER: Was it a much closer-run thing than you thought it would be?

MESSERVY: Yes. It was very difficult. Very difficult indeed.

REPORTER: Are these lads lucky to be alive?

There's a slight pause.

MESSERVY: No. Well ... No. They're all right.

For Doug Huntington, the scrum at Cork airport makes it difficult but not impossible to carry out his primary task, which is to make sure Harry Dempster didn't escape with the rolls of film, had he the mind too. 'My task was to keep an eye on the paparazzi and that this guy didn't escape. If he had got away with all the material it would have been serious.' Yet his concerns are ungrounded, as Dempster, ever the professional, sticks with the agreement, and his photographs, quickly developed and telexed around the world by the Press Association, illustrate the next day's newspapers.

The front page of the *Sunday Mail* declares 'Cheers', illustrated by a picture of Mallinson and Chapman being offered a bottle of champagne. The first line describes them as 'the jolliest Rogers in Britain', with Chapman quoted as saying, 'I feel great. It was very pleasant and comfortable on the bottom. The worst part was being brought to the top.' The *Sunday Mirror* declares: 'TOGETHER': 'The world held its breath – now they are safe ...'. (The rescue will eclipse the news that J. R. R. Tolkien, author of *The Hobbit* and *The Lord of the Rings*, had died at his home in Bournemouth, aged 81.)

In among the coverage is a brief interview with Doris Mallinson, Roger's mother: 'There were times when I thought

I would never see him alive again,' she said. 'I'm sure he will just want to forget about it and get plenty of sleep. He has not been in touch since the rescue, but thank God he's safe. That's all that matters now. Just to see him alive is everything.'

'We are pioneers,' says Messervy, 'and when you pioneer something you expect a problem or two.' On the subject of the global media attention he comments, 'You can kill 50,000 people a year on the roads but two men in a submarine and the world turns upside down.'

Both Chapman and Mallinson later give individual interviews to the national press. Mallinson speaks to the *Daily Mail* and testifies to the importance of Chapman in helping pull him through the darkest moments. The article has the feel, not of genuine quotes, but of sentiments strained through the newspaper's sieve to best resemble what they feel their readers wish to read. 'I owe my life today to Roger Chapman. The ex-Navy lieutenant, who was my second pilot and observer aboard the stricken mini-submarine *Pisces III*, pulled me through the blackest hours of that incredible rescue. Without him, I would not be here to tell the story.'

He later explains:

The carbon dioxide poisoning hit me badly. I had spasms when every joint in my body was racked with agonising pain. I also suffered blinding headaches. I know now that if it had not been for Roger I would have given up the ghost there and then. Words were not needed. But let me tell you how Roger brought me through. He held my hand and squeezed. It's true, and there's nothing odd or soppy about it. As we lay on two makeshift beds on either side of the circular, 6 ft, upside-down cabin that became an under-sea prison, that contact with another human being was enough to keep me going. There was no conscious decision to do it. It was just something that happened in the first few hours of our long, long wait and a tight squeeze of the hand expresses more concern than a thousand words.

In comparison, Roger Chapman's account, which appeared a few days later in the pages of the *Sunday Telegraph*, is more measured and detailed, but it too touches on their decision to share body heat:

> To start with, we had our two bunks separate, but we saw how ridiculous it was to lie there apart, both shivering, so we put the bunks side by side and lay there huddled together, front and back with our arms, round each other.

Chapman gives a brief glimpse of what they saw out of the thick glass portholes:

> The visibility was excellent – with our lights we could see about 15 feet – and there were a lot of fish about, mainly big members of the cod family, like saithe. Also we could hear a lot of dolphins: we never saw them, but their squeaks kept coming through our underwater telephone, and they interfered with our communications quite seriously.

When the *John Cabot* returned to Cork, the American team were more than in need of a little rest and recuperation. No sooner had the ship docked than Larry Brady gave one of the crew some money to buy a bottle of whiskey. Over the course of the return journey a slight feeling of resentment had begun to build that neither Chapman nor Mallinson had come on board to thank them for their efforts. As Brady would later write: 'There was no wasted time, no waving hands, only two tired submersible pilots ... who were swiftly loaded into the rubber boats and raced to their support ship. They never even looked our way.'

Yet after docking, and attracted by the promise of whiskey, a few of the men from Vickers Oceanics joined them, including one of the pilots. During the whiskey-drinking and mutual back-slapping, 'the Brits', as Brady described them, acknowledged 'that we had saved their mates' butts and asked, "How can we get hold of you guys again?"' In a moment of drunken bravado Brady pulled out his wallet, picked out a business card and handed it to him, telling him to stick it by his port-hole – 'just in case' – and to call anytime. (At roughly the same time Captain Len Edwards received an invitation to dinner from the BBC reporter Simon Dring, as the *Marina* had recently arrived in port. Edwards refused the invitation.)

After a brief sleep in the team's hotel in Cork, the party continued at an illegal after-hours drinking den. Rising at midnight and anxious for something to eat, Brady and the boys asked the night manager where they could go. He called a taxi and told them to tell the driver that 'Sean sent you'. After a long, disorientating drive through darkened streets, the taxi pulled up alongside a parked car, and the driver spoke to two men. Brady and the boys began to worry, only for them to be given the OK and ushered into a smoke-filled room, packed with illicit drinkers and diners.

The American team returned to San Diego on 5 September to a heroes' welcome. In recognition of their efforts, the city declared the day of their return 'CURV-III Day'. A press conference was held in the offices of the Port of San Diego and splashed on the front page of the next morning's *San Diego Union*. The 'tired but happy' eight-man crew were only too happy to share their adventures and take their rightful share of the credit for the rescue. Brady explained how the Canadians and British ships were hindered by 'tremendous seas', and that 'because of this, they waited for us to do it'. Of the toasted circuits that delayed their entrance by four hours, there was little mention.

The reunion at Walney was low-key. The gates to the small grass airfield were closed off and only a small group had gathered. Chapman stepped down off the plane and looked around for his wife. June was standing on the edge of the crowd, wearing a sheepskin coat. He walked towards her, smiling and casual, as if he'd just spent a weekend away. 'Hello, honey,' he said, to which June replied with equal calm, 'Hello, Rog, it's lovely to see you.' Then all pretence was set aside and they embraced. For Mallinson, the reunion with Pamela was emotional but restrained. He was relieved to see her and anxious to be reunited with the children, but all he wanted was to get away as quickly as possible, so they did not linger over the warm champagne in paper cups.

Messervy and Trice lingered a little longer. It was early evening, and Trice's wife and daughter were still staying with friends in Bath, so when Messervy invited him to stay over at his small rented cottage in Barrow, he accepted. Messervy's wife had never settled in the north and had left to return south. He was living alone, and they would eventually divorce. The house was cold when they arrived, but soon the fire was lit, large whiskies poured and the fight of both their lives was unspooled once again long into the night. In an interview with the BBC 25 years later, Messervy reflected poignantly, 'That was one of the good days of my life.'

The next morning was Sunday, and each gave thanks in their own way. Sir Leonard Redshaw rose early to pick gooseberries with his wife before taking to the skies. A champion glider pilot, after five days obsessing on the deepest depths he was anxious for the stillness and calm he found at 5,000 feet above the Cumbrian countryside. Mallinson stayed at home with his children, who were relieved at their father's return, while the Chapmans enjoyed the short walk to the village church where Roger had once read the lessons. As the minister reminded the congregation on this particular Sunday, there was so much for which to be thankful.

Peter Messervy was dismissive of any official inquiry. 'There's nothing an inquiry can glean that could not be learned over a couple of minute's chat round the table. I suppose we could look into the possibility of protecting the spindle with a type of safety guard. But it wouldn't need a big inquiry to recommend that. What happened was an act of God and no one should criticise us.' Yet an official inquiry there certainly was going to be.

The inquiry opened at 11 am on 5 September in the main conference room of Barrow Shipbuilding Works, attended by representatives from the Department of Trade and Industry, the American Bureau of Shipping, Hyco and the Royal Navy, Flag Officer Submarines. Both Messervy and Al Trice were among the 11-man panel. The panel was reconvened on 16 and 17 October 1973 to complete its investigation and establish whether any lessons could be learned.

The preliminary findings compiled by Greg Mott were telexed to J. G. Walmsley at the Department of Trade and Industry on 7 September 1973, with additional copies sent to the Ministry of Defence, Hyco in Canada and the American Bureau of Shipping. The report, just two pages long and consisting of 11 numbered paragraphs, stated:

In the course of this recovery operation, the hatch of the aft machinery compartment became dislodged. An eyewitness report indicated that the tow rope used in recovery became entangled in the locking mechanism of the hatch and possibly caused it to unlock.

A contributory cause may have been a small build-up of pressure in the aft sphere due to the ballast changes which took place during the surfacing of the submersible. This together with the motions of the submersible on the surface could have been sufficient to dislodge the hatch if the fouling by the rope had been such as to release the hatch locking mechanism.

In the sea state prevailing, dislodgement of the hatch could cause the compartment to flood and the submersible to sink.

The belief was that positive pressure inside the sphere pushed up on the hatch, so that, as soon as the rope loosened the bolt, the internal pressure blew off the hatch, which was now unrecoverable, buried as it was in silt on the seabed. The internal pressure could vary depending on the level of the hydraulic oil reservoir, the depth dived and the submarine's current trim and ballast.

The report stated that both *Pisces II* and *Pisces III* had been taken out of service until adequate alterations had been made to the locking arrangements on the hatch of the aft sphere, but that they were expected to resume service once adjustments were complete. Mott stated that he expected the final report to be completed by October, but it was not until the week before Christmas that the two-volume *Pisces III Accident Inquiry* was published.

The changes proposed caused an operational inconvenience. The current practice was to remove the hatch after every dive in order to check both the oil level in the aft sphere and whether there were any water leaks. The new proposals would mean completely removing the new aluminium cover plate prior to each inspection then replacing it immediately afterwards, adding a considerable delay to any turnaround. The longer-term solution was to reduce aft sphere checks to every tenth dive by fitting an improved oil level gauge, rendering a visual check redundant.

After the new oil gauge was fitted, the hatch was fitted with a transparent Perspex cover that was secured over the hatch and onto the main submersible by eight screws, protecting the locking hatch handle from any rogue ropes that might wash across in future. The fluke nature of the accident was corroborated by the fact that between them both *Pisces II* and *Pisces III* had already completed over 2,000 dives without any incident involving the sudden removal of the aft sphere.

In February 1973, five months before the accident, the latching mechanism that had been provided by Hyco was replaced by Vickers Oceanics with another mechanism, one also certified by the American Bureau of Shipping. It was largely identical, except that it was fitted with a new locking handle that acted like a spanner for turning the external hexagonal bolt to lock the hatch. The new handle, when locked, lined up with the vent plug (in theory, but not always in practice), allowing the vent plug cap to be screwed through a hole in the locking handle, which secured the handle in position.

Considered at the time to be an elegant piece of engineering, the new latch system achieved three goals: the main hexagon bolt was locked and now unable to rotate; the handle could not be secured without first fitting the vent plug through its handle; and, on surfacing, the removal of the vent plug to access the handle again would ensure that the internal and external pressures were equalised first before the hatch could be removed. However, the modifications meant the existing protective cover no longer fitted and so it was removed.

On the day of the accident there were further differences to the original Hyco system as a result of changes made by Vickers Oceanics. A different vent plug had been fitted and, most crucially, the locking handle was not fitted, which meant there was no longer the means to lock the latch in place. The reason for this was that on previous dives over the summer, water had been leaking into the aft sphere, so the seals around the shaft were replaced. The leaks, however, persisted. The vent plug was then replaced, but the remaining vent plug cap wouldn't work when used in conjunction with the locking handle. (The vent plug wouldn't sit properly, allowing small leaks to occur, which set off the bilge alarm.) The locking bar was removed six weeks prior to the accident.

Mallinson insists he repeatedly asked for the modifications to be made to the locking bar that would have brought it back into operation and had expected the repairs to have been made by the time he joined *Voyager*. As a consequence, *Pisces III* was in the habit of diving without any additional protection in place of the locked latching mechanism. This was now the operational 'new normal' over the summer months of 1973.

There was an additional contributing factor. The report concluded that at a depth of 1,575 feet, the outside water pressure pushing down on the hatch was the equivalent of 600 psi or 55 tons across the entire hatch. The report stated:

subjected to external pressure … the hatch can 'sink' or 'settle' into its seat and if this occurs the pins may come into contact with the lip of the hatch. It's possible, there-fore, that on returning to the surface the pins may not have been tight, in which case only a small torque would be necessary to rotate the shaft completely releasing the hatch mechanism … under such conditions any excess internal pressure would certainly facilitate the removal of the hatch by any external forces, or might even blow the hatch off, if the pressure was sufficient. A pressure differential of less than 1 lb/sq. in. would be sufficient to lift the hatch and the pressure could easily be generated by return of the oil to the aft sphere.

The aft sphere is a steel pressure-tight sphere divided into an upper and lower compartment by a wooden floor. The lower compartment contains the hydraulic oil reservoir and the upper compartment houses the hydraulic pump motor, an inverter and other small items of equipment. If water penetrates the access hatch, warning lights flash in the main sphere and an alarm will sound.

During the inquiry, the members of the panel stated that the aluminium fibreglass fairing had been removed, but when interviewed Mallinson corrected them, explaining that it had never been fitted. When Chapman was asked the same question he backed up Mallinson's statement, and when the panel insisted that he was speaking about events that might have preceded his appointment to the company, Chapman coolly pointed out that an inspection of the sub would reveal that no holes had been drilled to take the fairing. 'I was forever grateful to Roger for backing me up,' Mallinson later recalled.

As part of the inquiry, Chapman and Mallinson also prepared a short report detailing their experiences, accompanied by recommendations. Under the category of 'food and drink', they stated that the 'emergency ration pack was adequate and although it's recommended that water be included in the future, condensation provided enough water on that occasion'. How an emergency ration pack that neglected to include water, leaving both men to survive for more than three days on condensation lapped from the steel frame, could even be described as 'adequate' is an indication of the extreme latitude Vickers Oceanics decided to afford themselves. On 'hygiene', the two men reported: 'The pilots managed reasonably with the normal sanitary bottle and plastic bags, but recommended that extra provisions should be standardised.'

On oxygen and carbon dioxide they stated: 'Oxygen consumption during the 8 hour working dive was 0.382 litres/min/man, and 0.3 litres/min/man during the 75½ hour rescue period.'

The report stated that the men could survive until 12 noon on Saturday 1 September and were rescued at 1.17 pm on the Saturday, but that 'life support proved to be somewhat greater than was expected from original estimates'.

The report also touched on the suggestions submitted by members of the public gripped by the rescue.

Many suggestions have been received from the public about plugging in an air hose from the surface to lengthen the life-support time. The existence of a suitable flexible hose capable of withstanding the external pressures encountered is not known to VOL. Also the problems of a ship in a seaway maintaining an accurate position over the submersible in order to supply air are not fully understood by the public.

The recommendation is that life-support supplies be increased from 72 hours to 100 hours, exposure suits fitted, alongside increased supplies of food and water and heavier plastic bags for 'waste retention'.

The inquiry found that 'a principal factor' in the cause of the accident was the decision made by 'field personnel' on or about 15 July 1973 not to replace the locking bar on the hatch securing arrangement on the aft sphere. The report stated:

> This was done because of the difficulties experienced in preventing a slight leakage through the spindle seal of the securing mechanism and the vent valve, which gave rise to frequent operation of the flooding alarm. The vent valve was also being damaged by the locking bar. The Company procedures do not allow modifications to the submersible affecting its safety to be carried out by field personnel without the prior written approval of the Technical Manager. Approval was not sought nor given for this modification.

Among the recommendations of the inquiry was that further research was required into how the press should be handled.

> The physical interference with the rescue operation by a trawler on charter to the Press, and not playing a useful role in the operations were discussed. The interference caused to underwater communications and tracking by propeller noise of such vessels was particularly serious. Requests by VHF, signalling lamp and a notice board carried by a helicopter had been deliberately disregarded ... Responsible Press authorities need to be told in no uncertain terms of the problems which their representatives can cause in their quest for news. The matter is one of

substance worldwide and the Inquiry Panel recommended attention by those concerned with developing law applicable offshore. Court proceedings taken after the event are no substitute for sensible actions during rescue operations.

The exact amount of oxygen that remained within *Pisces III* still remains a matter of debate. Hours after their rescue, both Mallinson and Chapman said there was at least 24 hours – or a day's supply – left. This may or may not have been as a result of pressure from Vickers to maintain an image of relaxed efficiency. Yet in his memoir *No Time on Our Side*, Roger Chapman suggests that there might have been less than 20 minutes left.

Doug Huntington remembers climbing inside *Pisces III*, accompanied by an inspector from the American Bureau of Shipping, when the vessel was brought ashore in Cork. He remembers two things about what he discovered when they went in. The first was the stench: 'It was awful.' The second was the oxygen supply. Huntington remembers the inspector opening the valve on the last remaining oxygen tank: 'There was a little puff of O_2 which came out, and that was it. It was the last gasp. They were right at the end. They had nothing left. They were about to go to sleep.'

A year to the day after the rescue, the phone rang in Broom House at precisely 1.17 pm. Roger Chapman smiled before he picked the receiver off the cradle. He knew exactly who it would be.

EPILOGUE

For six centuries the Church House Inn has welcomed the weary and the footsore, the hungry and the thirsty. Sitting in the hills above Coniston Water, nestled on the southern fringes of the Lake District, the two-storey inn has white-washed stone walls with black painted window lintels. Outside, a beer garden echoes to the sound of cheery travellers, while inside there's a bar propped up by oak ale casks, a ceiling of low wooden beams and a comforting fire of split wood, crackling and spitting in the grate. There are few places better for an annual reunion, and certainly a good many worse. Equidistant from their two homes, it was here on 1 September each year that Roger Chapman and Roger Mallinson would meet to mark their survival.

It began with that phone call, at precisely 1.17 pm on 1 September 1974. The call was to become a yearly ritual, ringing out over the next 40 or so years and marking the exact moment of their surfacing. Over the decades the calls marked new careers and, for Chapman, new children: two boys, Marcus and Sam. Then there were life's upsets and disappointments, troubles and delights. By the time of the first telephone call, both men had already returned to the deep on different dives on the *Pisces*-class submersible, but Chapman would be the first to surface and move on to new pastures in the field of accountancy. But the waters of the deep would once again sound her siren call.

Mallinson would often be the first to arrive for lunch at the Church House Inn, as it was only a 30-minute drive from Windermere, where he now lived alone in a little cottage crammed with photographs and paintings, a massive gramophone trumpet speaker built by his brother Miles, and against one wall an impressive pipe organ. On discovering that the cottage was too small for any commercially available organ, Mallinson designed and built a bespoke one for himself.

He would never lose his passion for engineering and nautical pursuits. In 1976, on discovering that the SL *Shamrock*, one of the historic old steam vessels that plied its trade on Windermere, was set for destruction, he bought it and set about restoring it from a state of dilapidation to its original steam power. Today it can still be seen on the lake and is a popular tourist attraction, one for which Mallinson was awarded a Lifetime Achievement Award by the Transport Trust. Mallinson's mode of transport to the Church House Inn remained his trusty mint-green 1931 Austin 7, bought back in 1957 and now with over one million miles on the clock. His co-pilot, sitting in the passenger seat, and with his own pair of orange waterproofs against inclement showers, was often his German shepherd Whappet, whom Mallinson had actually trained to steer the car at classic car rallies, while he worked the gears. At the Church House Inn, he favoured one of the wooden tables near the fire, close enough to the bar to easily catch the barman's eye and ensure a swift order.

The *Pisces III* accident would remain a source of annoyance to Mallinson. He could never bring himself to make peace with the official inquiry and believes to this day that there was something more to be unearthed, hidden behind a conspiracy of silence. He believes the hatch was incorrectly connected to the tow line and it was the pull of the ship that helped to yank it off, rather than the million-to-one chance of a rope wrapped around a nut, assisted by a build-up of pressure in the aft sphere. To those who ask, he describes the official report as 'rubbish', and laments that expending so much money and effort was necessary to save their lives and fix a crisis that should never have happened in the first place.

Roger Chapman, meanwhile, had moved away from the little village of Broughton, and would use his personal experiences in the depths of the Atlantic to both launch a successful business and save lives. In the 1970s, after that brief dalliance with the world of accountancy, he returned to the sea and set up Sub Sea Surveys Ltd to take advantage of the financial boom triggered by the discovery of oil in the North Sea. The company would take inspiration from the American CURV system and bring the first all-electric remote-operated underwater vehicle to Britain, where it worked with oil companies installing pipelines. In time, Vickers Oceanics would buy it, and following a spell working for his old company, Chapman would go on to launch Rumic, a new company based in his native Cumbria, where he would pioneer a new generation of both manned and unmanned submersibles designed to assist in the oil industry and in subsea rescues for the military.

Chapman would never forget the fear and terror of those three and a half days in the darkness of the deep. He developed the *LR5*, a manned miniature submarine designed to dock at depth with military submarines that was capable of rescuing 16 men at a time from a stricken vessel.

In August 2000, while on holiday in Umbria with June and his sons Marcus and Sam, Chapman received a call from Rumic's head office in Cumbria with news that the Royal Navy were despatching the *LR5* and Rumic's rescue team to the Barents Sea. June drove him to Rome airport, where he caught the next flight back to Britain. The next ten days were to be among the most frustrating and depressing of his career. The incident in the Barents Sea involved the sinking of the Russian nuclear submarine *Kursk* after an explosion of torpedo warheads during a training exercise. For five days the Russian government would refuse all offers of international assistance, and by the time the LR5 arrived at the scene on board the Norwegian ship *Normand Pioneer* it was too late: all 118 crew on the *Kursk* were dead. Privately Chapman would blame not only the intransigence of the Russian Navy and President Vladimir Putin, but also the Royal Navy for spooking the Russians.

Yet it was to be a different story five years later. In August 2005 the *AS-28*, a 44-foot Russian rescue submersible with seven naval crew on board, became trapped 600 feet down in the waters off the Kamchatka Peninsula in the Russian Far East when its propellers got entangled in cables that were part of Russia's coastal-monitoring system. In 2000 President Putin had been criticised by the nation's press for continuing his holiday during the *Kursk* disaster, but this time the

Russian government was swift to seek international assistance. Over 30 years on from the rescue of *Pisces III*, Chapman found himself cast in the role of Commander Peter Messervy as he helped to coordinate rescue forces in another race against time to save men from the deep.

A Scorpio 45 remotely operated vehicle with support equipment and crew were swiftly mobilised to Prestwick airport in Scotland and fitted on board an RAF C-17 transport plane, which flew 4,600 miles to the Kamchatka Peninsula. While waiting for international assistance, the Russian Navy had first attempted to lift the sub, and when this failed they attempted to drag it into shallower waters. This was also unsuccessful, as the sub was held by cables secured in place by a 50-ton anchor. Inside the submersible the seven crewmen huddled together for warmth, minimising their movements in order to conserve oxygen. When finally launched, the Scorpio remote-controlled vehicle was able to cut the cables that trapped the Russian submersible, enabling it to surface with only a few hours of oxygen left. During a visit to Britain for an EU–Russia summit, Vladimir Putin met the team from the Royal Navy and Rumic – which had since been bought by James Fisher and Sons – at Downing Street. Chapman described the president of Russia as the 'shortest, scariest man' he'd ever met.

When Roger Chapman comes through the door of the Church House Inn he's looking older. The receding hair of his late 20s has long since beaten a full retreat and his face is lined with age, but his eyes are timeless, with a mix of mischief and intelligence. Mallinson looks up and smiles, and when

Chapman asks how he is, the older man replies with a comic sigh: 'Still here.' They both are, and this annual lunch marks the fact that they might so easily not have been. Looking back, there were so many twists of fate, so many elements in the rescue that went wrong, problems that spiralled and spiralled, eating into their remaining time and oxygen supply. What if the rescue had simply been too late? It doesn't bear thinking about, although if they're honest, on occasion both men do.

Chapman takes a seat by the fire. Summer is still here, but the wood sparks and spits in the grate. Today there will be pale ale instead of stale lemonade, gammon and chips instead of emergency rations, and, they joke, all the air they can breathe. There will be recollections, reminiscences and regrets, anecdotes and jokes. Although their lives – hard won – have gone in two different directions, there remains an unbreakable connection.

For in their memories, a part of each man is forever in the dark of the deep, together alone.

NOTES ON SOURCES

This account of the sinking and rescue of *Pisces III* is based on original interviews, contemporary newspaper accounts and official documentation. I've had access to the official report, *Pisces III Accident Inquiry* vols I and II, published on 20 December 1973, which contains the pilot's log, extracts from the log kept at the Vickers base in both Barrow and Cork, as well as assorted correspondence and technical reports. 'The Sinking and Rescue of *Pisces III*' by Harold Pass, a paper presented at the Seventh Undersea Medical Society Workshop in San Diego in November 1974, was collected in a volume by the National Technical Information Service, US Department of Commerce.

The two primary voices in the book belong to Roger Chapman, who sadly died in January 2020, and Roger Mallinson. I was fortunate to conduct extensive interviews with Roger Mallinson, while Roger Chapman documented his experience in *No Time on Our Side* (W. W. Norton & Company, 1975). June Chapman was generous in her support and also gave me an interview with her own recollections, as did a number of key individuals involved in the rescue from Britain, Canada and the United States. These include from

Vickers: David Mayo, Doug Huntington, Ted Carter, Maurice Byham, Terry Storey, Bob Hanley and Dick Nesbitt; from Hyco I interviewed Jim McBeth, Mike Macdonald and Al Trice; and from the American rescue team I interviewed Larry Brady and Bob Watts.

The Canadian contribution to the rescue was also documented in a detailed but previously unseen Hyco memo, 'Canadian Contribution to the *No Time on Our Side* Story', dated 1 February 2000. The American involvement is documented in a short memoir, 'It's Just a Matter of Time' by Robert Wernli and Larry Brady. An alternative source is the 'The U.S. Navy Participation in the rescue of Pisces III' by Howard R. Talkington, published in January 1974 in the *Marine Technological Society Journal*, Vol. 8. No. 1.

Material on the history of the submarine was drawn from *The Submarine Pioneers* by Richard Compton-Hall (Sutton Publishing Limited, 1999) and *The Deadly Trade: The Complete History of Submarine Warfare from Archimedes to the Present Day* by Iain Ballantyne (Weidenfeld & Nicolson, 2018). The history of the original transatlantic telecommunications cable was drawn from John Steele Gordon's excellent book *A Thread Across the Ocean* (Harper Perennial, 2003) and *The Cable: Wire to the New World* by Gillian Cookson (The History Press, 2012). For the background history of Vickers I am indebted to two books in particular: *Vickers Against the Odds 1956/77* by Harold Evans (Hodder & Stoughton, 1978) and *Vickers' Master Shipbuilder Sir Leonard Redshaw* by Leslie M. Shore (Black Dwarf Publications, 2011). The creation of the *Pisces* submarine is drawn from *Westcoasters: Boats That Built BC* by Tom

Henry (Harbour Publishing, 1998). Other books to which I referred included *Illustrated Encyclopedia of the Ocean* (Dorling Kindersley, 2011), *A History of Modern Britain* by Andrew Marr (Macmillan, 2007), *State of Emergency: Britain 1970–1974* by Dominic Sandbrook (Allen Lane, 2010) and *The Body: A Guide for Occupants* by Bill Bryson (Doubleday, 2019). Additional information was drawn from contemporary reports in *The Times*, *Daily Mail*, *Sunday Mirror*, *Daily Express*, *Daily Telegraph*, *San Diego Union* and *New Scientist*.

ACKNOWLEDGEMENTS

I will not be the only author to have faced the challenges of researching and writing a book in the midst of a global pandemic, but the predicament was made easier by the support and assistance of so many people who believed this largely forgotten tale deserved to be retold. I would like to start by thanking the two people alive today who back in 1973 had the most to lose: Roger Mallinson, who gave generously of his time for a series of interviews, as did June Chapman, the wife of Roger Chapman, who sadly died before I had the opportunity to speak to him. June and the couple's sons Sam and Marcus were hugely supportive, providing me with generous access to Roger's records and files.

Among the former staff of Vickers Oceanics I would like to thank the following for their time and support: Doug Huntington, Ted Carter, Maurice Byham, Terry Storey, Bob Hanley, David Mayo, Dick Nesbitt and Des D'Arcy. In Canada I would like to thank Al Trice, Jim McBeth and Mike Macdonald, as well as Vickie Jensen, who is writing her own book on the history of Hyco, and yet could not have been more generous with her support and assistance on this project. In the US I am indebted to the assistance of Larry

345

Brady and Bob Watts, as well as Robert Wernli and Tom Lapuzza.

At HarperCollins I was superbly assisted by a dedicated team led by my editor Joel Simons, including copy editor Mark Bolland and project editor Sarah Hammond, as well as Fiona Greenway, Fionnuala Barrett and Ameena Ghori-Khan. My first book was published by HarperCollins in 2003, and, almost twenty years later, it's good to be back and see that their support, dedication and professionalism remain unchanged.

I would especially like to thank James Spackman and Jason Bartholomew, my agents at the BKS Agency, for unstinting support during what, at times, was a challenging year, and I would also like to thank Sarah Brown for making the first introduction. Mark Gordon and Beth Pattinson at Mark Gordon Pictures have also shown great enthusiasm for the project beyond the printed page, and I appreciate their continued support. I would also like to thank Dr Stephen Hearns for reading over the medical sections. Many people have assisted me in my attempts to get things right, but if errors do occur the responsibility is mine alone.

Last in the acknowledgements – but always first in my life – I would like to thank, from the depths of my heart, my wife Lori.

INDEX